建设工程管理中英文双语丛书
Construction Management Series

市场经济下的建设管理
Construction Management in a Market Economy

Denny McGeorge Angela Palmer Patrick X.W.Zou
丹尼·麦乔治 安琪拉·保玛 邹小伟

中国建筑工业出版社

著作权合同登记图字:01-2002-4979号

图书在版编目(CIP)数据

市场经济下的建设管理/(英)麦乔治,(英)保玛编著;邹小伟等译.—北京:中国建筑工业出版,2003
(建设工程管理中英文双语丛书)

ISBN 7-112-06028-1

Ⅰ.市… Ⅱ.①麦…②保…③邹… Ⅲ.建筑业—经济管理—汉、英 Ⅳ.F407.9

中国版本国书馆 CIP 数据核字(2003)第 083516 号

Copyright© 1997 Blackwell Science

All rights reserved. No part of this publication may be reproduced, stored in a retrieval system, or transmitted in any form or by any means, electronic, mechanical, photocopying, recording or otherwise, except as permitted by the UK Copyright, Designs and Patents Act 1988, without prior pernlission of the publisher.

本书由 Blackwell Science 出版公司正式授权我社在中国翻译、出版、发行
中英文版

责任编辑:常 燕

市场经济下的建设管理
丹尼·麦乔治
安琪拉·保玛 著
邹 小 伟 译

中国建筑工业出版社出版、发行(北京西郊百万庄)
新 华 书 店 经 销
广东省肇庆市科建印刷有限公司

开本:787×1092 毫米 1/16 印张:21¾
2003 年 10 月第一版 2003 年 10 月第一次印刷
定价:38.00 元

ISBN 7-112-06028-1
F·482(12041)

版权所有 翻印必究
如有印装质量问题,可寄本社退换
(邮政编码 100037)

本社网址:http://www.china-abp.com.cn
网上书店:http://www.china-building.com.cn

序

得知我校建筑环境学院的丹尼·麦乔治教授、邹小伟博士与在英国的安琪拉·保玛博士合作，编写了这本有关建设管理的书，并将以中英文对照的形式出版，我非常高兴。本书的三位作者都具有相当丰富的在亚洲工作的经验。

我们大学很高兴曾与中国的朋友、同行有着广泛的专业技术合作。这种合作不仅是在专业技术领域，同时也包括政府间关系及工商界企业间的商务活动等方面。这种日益增长的合作关系是建立在中国及澳大利亚两国各自所拥有的专业技术及经验基础上的。也因而使一些特别的技能可以在中澳两国间共享。

本书是为建设管理专业的学生及该领域的实践工作者而编著的。它也将有助于中澳两国间在该领域的思想交流及研究工作的发展。

我校为此书的成功出版向本书的作者及建筑环境学院表示真诚的祝贺。

余森美
新南威尔士大学 校长
澳大利亚 悉尼

前　言

《市场经济下的建设管理》这本书的写作目的，是想给中国的读者介绍西方国家在基本建设过程中所运用的一些新的管理方法。本书所提到的建设过程包括从建设的最开始的设想阶段，到完全建成可以交付使用的一系列过程。

本书的三位作者——荣誉教授 丹尼·麦乔治（Emeritus Professor Denny McGeorge）、安琪拉·保玛博士（Dr Angela Palmer）和邹小伟博士（Dr Patrick X.W.Zou）在中西方建设管理方面都有丰富的经验。希望这些经验能够满足中国读者的需求。我们不能冒昧地说这是一本指导读者在中国进行建设管理的手册。本书旨在介绍目前西方国家在建设管理中正在使用的、对中国读者有用的方法。

本书以论题的形式，集主要的管理新技术：目标管理（Benchmarking）、重组（Reengineering）、合伙人制（Partnering）、价值管理（Value Management）、可建设性（Constructability）和全面质量管理（Total Quality Management）于一册。在最后一章"中西方建筑行业中的建设管理的当前议题"中，我们讨论了一些中国建筑业目前所关注的几个问题。

本书客观地介绍了有关建设管理的概念和这些概念之间的相互关系。本书内容适合于建设管理专业的研究生及高年级本科生的学习参考，也适用于从事建筑业的管理人员。

在此值得一提的是，在西方建筑管理方面，并没有绝对统一的方法。比如，当英国的建筑公司想要用美国的价值工程系统（Value engineering）时，他们发现这行不通。这里面原因很多，但主要原因在于价值工程系统研究的最初目标。美国价值工程系统源于政府对工程巨大责任的需求，在美国几乎所有价值工程所做的工作都是政府工作。而在英国，情况就非常不同。英国的量数调查系统（Quantity survey）（成本控制系统）规定了所必须的一切责任。价值工程应该提供一个用来检验成本与价值关系的平台。这个例子说明在西方国家之间由于文化的差异对于同一管理概念存在着不同认识和不同做法。在重组（Reengineering）的方法上，美国和欧洲之间也同样存在着差异。

可以预料，中国也将发展自己的建设管理概念。比如中国自己的重组（Reengneering）、全面质量管理（Total Quality Management）、目标管理（Benchmarking）、价值管理（Value Management）。然而，很有可能由于文化的差异，西方一些观念会不适合中国的实践。而我们并没有把那些可能不适合中国建设管理的概念或管理技术分离出来。这是因为，鉴于中国当前处于从计划经济向市场经济转化的过程当中，这一大环境是影响建筑行业的一个重要因素。在此情况下，即使在长期实践中，某些概念并没有被采用，但我们认为在初始阶段，有必要把各种概念都给予阐述。

另外需要指出，中国建筑业也正在发展自己的建设形式。这包括有形建筑市场的建立。本书在最后一章对此将有详细的介绍。

为了方便中国的读者，我们将本书作了中文翻译。中国刚加入世界贸易组织（WTO），这意味着借鉴西方的管理技术是很必要的，它能促使中国在这一方面的发展更具有可持续性的竞争力。我们真诚地希望，我们认真编写的这本书能对从事建设工程项目管理及研究工作的广大读者有所帮助。

丹尼·麦乔治　　荣誉教授
安琪拉·保玛　　博士
邹小伟　　　　　博士

二○○三年 六月
澳大利亚·悉尼

目 录

序
前言
第一章 绪 论 ·· 1
　本书内容 ·· 2
第二章 目标管理(Benchmarking) ··· 4
　导 言 ·· 4
　目标管理的定义 ·· 5
　目标管理发展史 ·· 6
　目标管理的类型 ·· 6
　目标管理的过程 ·· 9
　目标管理工作组 ··· 19
　目标管理行为准则 ·· 19
　法律条文的规定 ··· 20
　目标管理:主要议题 ·· 21
　当前的研究 ··· 22
　案例分析 ·· 23
　结 论 ·· 24
第三章 重 组(Reengineering) ·· 26
　导 言 ·· 26
　重组的起源 ··· 27
　建筑业中的重组 ··· 27
　重组的目标 ··· 28
　重组方法论 ··· 30
　重组的陷阱 ··· 33
　信息技术和重组 ··· 35
　欧洲人对重组的看法 ··· 37
　一个澳大利亚建筑业的过程重组研究的案例分析 ···························· 37
　结 论 ·· 45
第四章 合伙人制(Partnering) ·· 47
　导 言 ·· 47
　合伙人制的起源 ··· 47
　建筑业内的合伙人制 ··· 48
　合伙人制的目的 ··· 48
　合伙人制类型 ·· 49

项目合伙人制 ··· 49
　　策略或多项目合伙人制 ··· 56
　　合伙人制在法律和合同上的意义 ·· 58
　　争议解决 ·· 60
　　结　论 ·· 60

第五章　价值管理（Value Management）·· 63
　　导　言 ·· 63
　　发展历史 ·· 63
　　功能分析 ·· 66
　　价值管理研究的组织 ··· 72
　　应该由谁来实施研究？ ·· 73
　　团队该由哪些人员组成？ ··· 73
　　价值管理研究形式 ·· 75
　　应在哪里开展研究？ ··· 75
　　研究开始的时间 ·· 75
　　如何评估可选方案 ·· 76
　　价值管理作为一个系统 ·· 77
　　美国价值工程系统 ·· 78
　　一个美国的价值管理案例分析 ··· 79
　　英国价值管理系统 ·· 80
　　一个英国的价值管理案例分析 ··· 81
　　日本的价值管理系统 ··· 83
　　一个日本的价值管理案例分析 ··· 83
　　为什么这些系统互不相同？ ·· 84
　　价值管理与造价工程间的关系 ··· 85
　　结　论 ·· 86

第六章　可建设性（Constructability）·· 88
　　导　言 ·· 88
　　起　源 ·· 88
　　可建设性的目标 ·· 89
　　实施可建设性 ··· 91
　　实践中的可建设性 ·· 92
　　可建设性和楼宇产品 ··· 94
　　好的与坏的可建设性 ··· 97
　　将可建设性的利益量化 ··· 100
　　结　论 ·· 101

第七章　全面质量管理 ·· 102
　　导　言 ·· 102
　　全面质量管理的定义 ··· 103

什么是质量？	104
全面质量管理的发展历史	106
模式转变的需要	108
建筑业的文化转变	108
以顾客为中心	110
整 合	110
全面质量管理的包容性	115
不断提高	115
质量成本和质量所造成的成本	115
国际质量标准，例如 ISO 9000	116
改变管理	117
如何实施全面质量管理	117
凯审（Kaizen）	118
目前对建筑业全面质量管理的研究	118
结 论	119

第八章 西方与中国建筑业中建设管理的当前问题 ... 120

导 言	120
文化趋势	121
现行的议题	122
新出现的问题	124
未来方向	125
结 论	127

第一章

绪　　论

目前的建筑行业对提高质量,控制成本和减少合同纠纷等方面有着强烈的需求。于是许多新的管理技术被采纳,试图用来解决这些问题。但行内许多人发现这些概念都比较模糊,因而怀疑这些概念的实用性。

我们编著此书出于两个目的。首先,我们认为有必要首次把目前在建筑工业中提倡的诸多管理概念集于一册,并逐章加以介绍。我们所选择并准备加以介绍的概念包括:目标管理(Renchmarking)、重组(Reengineering)、合伙人制(Partnering)、价值管理(Value management)、可建性(Constructability)及全面质量管理(Total quality management)。虽然我们不敢承诺这本书包含了建筑领域的所有内容,但它的覆盖面非常广,这也正是本书的第二目的,即如何面临应用不同的管理概念而取得协调的成果这一挑战。

在同本行业工作者的交谈和接触中我们意识到,对于现代管理概念的实效存在着许多有意义的质疑。很多建筑业者都认同管理概念经过六个阶段的看法。他们是:以前概念的不足;发现或重新发现一个解决方法;为成功的发布而喜悦;过多用于不恰当的场合;由于失败例子的增多而嘲笑或忽视该概念;由于该概念被抛弃或是被新的概念代替以至于最后放弃该概念。尽管以上的看法并不是我们自己的,我们的确认为将真正的工业文化的转变从一些时尚的想法中区分出来是很重要的。在这一点上,戈得夫列(Godfrey)也曾作出警告:"合伙人制关系发展得很快,但也有一个危险,那就是它也将会流行一时而后消失。"

本书写作的背景是实践工作者们已经具有了许多管理概念。虽然一些概念,比如全面质量管理、价值管理和目标管理是无可争议的,但一个实在的危险就是工业界人士对于过去十年来所出现的众多管理概念,以及这些概念之间的相互更迭,使得他们无计可施,而且失望。近期在香港的一次调查中发现,很多建筑业的客户以为,"价值工程只不过是节省成本和可建设性研究的代名词,或者是成本控制中的一种技术"。而对于每一个新的管理概念,其倡导者都会极力推崇"他们的"技术而排斥其他技术。因而使这种饱和情绪和失望更甚。因此从实际工作者的角度来看,一些概念似乎是相互排斥的而不是相互补充的,而且每一个新的概念都在争先恐后地吸引决策者的注意力。这种积极推崇某一概念的现象,用翰默(Hammer)的关于重组的口号来说便是"不要自动化,使之消失",或者用科拉达(Kelada)带有修辞色彩的疑问便是"重组是否正在取代全面质量管理"。

在本书当中,我们尽量给所挑选的概念以直接和客观的说明。本书的目的是让本行业工作者和学术研究者了解建设管理概念的发展水平和状况,同时作者提出了一个概念化的模式。其目的是为了使读者能够更好的理解这些概念和那些有待发展的概念之间的内在关系。

本书自始至终我们都在努力区分韦伯字典(Webster Dictionary)中概念和技术的区别。前者是由一些特定的例子总结出的抽象概念,后者则被定义为那些为了实现一个目标而使用的方法。本书既涉及到概念，也涉及到技术方法。理性的学习应是概念化的。

本书内容

第二章：目标管理(Benchmarking)

　　目标管理是一个以通过检验和提炼经营过程来提高一个组织的竞争力为目标的概念。这一概念源于施乐公司(Xeror)，该公司将对手的复印机拆卸后，与本公司的复印机进行详细比较。后来，该公司将这种比较扩展到整个商业过程。本章涉及到：目标管理的种类；目标管理的过程；目标管理小组；目标管理准则。本章的结尾列举了一个简单实例，一家全国性的房屋建造商就以顾客为中心的目标管理将其商业过程与一家全国性的汽车制造商的商业过程进行比较。

第三章：重组(Reengineering)

　　重组被认为是管理的革新，它可以和亚当·史密斯(Adam Smith)的《民族的财富》("Wealth of Nations")之中提到的工业革命相提并论。自从它的引进，倡导企业经营过程重组已取得了许多显著的成果。虽然人们还没有完全理解重组在建筑业的影响，但人们对这方面的兴趣已经越来越浓。本书包含重组的以下方面：起源；建筑业的重组；目的；方法论；实施；时间和成本的节约；重组的缺陷；信息技术和重组的关系；欧洲在重组方面的概况。本章还包括了一个详细的工程实例，即T40工程。这是澳大利亚建筑工业过程重组的一个实例，包括从计划到执行的整个过程。这一项目的目标是使建造时间减少40%。

第四章：合伙人制(Partnering)

　　合伙人制正式的概念的起源并不远，可以追溯到20世纪80年代中期。它起源于美国，之后传播到其他国家，包括南半球的澳大利亚和新西兰，也传播到了英国。合伙双方采用合伙方式使传统的对立关系转变为"双赢"情形。合伙可以是在一个独立的项目上的短时期的合伙，也可以是战略性的长期合伙。本章将描述：合伙制度的起源；建筑业中的合伙；合伙的目标；合伙的种类，项目合伙和战略合伙；合伙参与方各方所承担的义务和许诺；合伙的过程；如何组织合伙讨论会；合伙宪章；合伙的缺陷；合伙的局限；合伙的法律规定和合同含义；合伙纠纷的解决。本章结尾对整体建筑业在合伙制方面的实施进行了全面推测。

第五章：价值管理(Value Management)

　　价值管理源于第二次世界大战期间美国的制造业。其目的是通过加强产品的功能来提高其价值。价值管理在制造业管理中非常成功，因此美国国防部将它用于建筑业的管理。大约与此同时，英国的建筑业也表示了他们对价值管理的兴趣。本章追溯：价值管理发展史；功能分析的应用；价值管理研究的组织；价值管理提案的评估；美国价值管理系统；英国价值管理系统；日本价值管理系统。本章最后分析这三个价值管理系统的区别，并考察了主要的文化对于价值管理发展的影响。

第六章：可建设性（Constructability）

可建设性是本书中惟一的一个专属建筑业领域的管理概念。可建设性是关于如何使建造过程中的决策有助于提高建造的质量。可建设性概念起源于20世纪80年代初，从最初很局限的应用范围发展到公司决策支持理论和系统。本书涵盖了可建设性的以下几个方面：起源；范围和目的；实施；在实践中的可建设性；使用过程中的建筑物；好的和不好的可建设性——成功的指标；可建设性好处的计量。本章最后以区分可建设性和良好的多学科团队协作而结束。

第七章：全面质量管理（Total Quality Management）

全面质量管理是通过对以客户为中心的方法，组织过程的整合，不断提高的原理的加强，来促进组织发展的概念。本章探讨全面质量管理的概念；发展历史；建筑业中文化改变的需要；以客户为中心；整合；不断提高；质量成本和质量标准。本章最后简要的分析目前正在使用的一系列质量方法。

第八章：中西方建筑业中的建设管理的当前问题（Current Construction Management Issues in Westen and Chinese Construction Industries）

这一章正如它的标题所示，主要的信息来源于西方的市场经济体制，如在英国和澳大利亚的市场经济下，政府的主要角色是通过管理实践的加强和提高来促使工业文化转变。在西方经济体制下，它导致了对一些管理技术的采用如目标管理，全面质量管理，可建设性，价值管理，合伙人制和重组。在中国，也有与之相对应的首创，比如有形建筑市场。这一章中我们提出了对当前形势的一种看法，即越来越多的西方管理概念都有潜在的问题。设施管理的增长以及设计和建造专业人员同客户之间关系的变化可以用帕累托（Pareto）影响曲线来表示。我们认为全球化过程中不可预测的社会、经济、技术和政治等各方面的因素将迫使各种公司组织考虑组成国内的和国际的商业联盟。工程项目联盟是未来十年里越来越受欢迎的管理实践。本章最后还提出了中西方加强交流的必要性，以及通过相互学习对各自都有益处的观点。

正如前言中所提，值得强调的是，西方的建设管理也并不是统一的。比如当英国的建设公司想要采用美国的价值工程系统时，他们发现行不通。这里面原因很多，但主要原因关系到价值工程研究的最初目标。美国价值工程系统源于政府对工程项目的必需的巨大义务和责任。在美国几乎所有价值工程所做的工作都是政府工作。而在英国，情况就非常不同。英国的量数调查系统（Quantity Surveying）（成本控制系统）已对必要的义务和责任做了规定。美国的价值工程是用来提供一个检验价值与造价关系的平台。这个例子说明在西方国家之间由于文化的差异造成的对于同一管理概念的不同认识和不同做法。另外，在重组上，美国的方法和欧洲的方法之间也存在着差异。

可以预料，中国也将发展自己的建设管理概念。比如重组、全面质量管理、目标管理和价值管理。很有可能由于文化的差异，一些西方概念会不适用于中国的实际情况，而我们并没有把那些不适用于中国建筑业的管理概念或技术分离出来。考虑到中国建筑业正在从计划经济向市场经济转化的过程中，我们相信，即使有一些概念在一段时间后将不会被采纳，我们还是应该在最初阶段将所有的概念都考虑到。

第二章

目标管理（Benchmarking）

导 言

米歇尔·史密斯（Michelle Smith），一位爱尔兰游泳健将，于1996年亚特兰大奥运会上一举夺得三枚金牌，震惊了整个泳坛。当被问及成功的秘诀时，她说，除了其他一些事情，她学习了田径运动员的训练方法，并将其运用到她的游泳训练计划中去。米歇尔的训练方法并不是她创新的。许多运动员也都曾从其他运动项目中寻找那些成功的训练计划中的新技术。依玫·扎托派克（Emil Zatopeck）是唯一一位在同一届奥运会中获得三项长跑金牌的选手，他从自己的军队生涯中学习训练的技术。其他运动员，例如罗·西尔（Ron Hill），世界著名马拉松运动员之一，采用瑞典生物学家发明的碳水化合物饮食法来提高他的成绩。据说有的运动员采用的食谱中还包括海龟血和沉淀的犀牛角。

所有这些经验的共同点是从本专业范围以外的领域中寻找改进的方法。他们使用了其自身运动项目中广泛被采用的训练方法，但这些还不够，因为每个人都在使用这些方法。为了真正成功，他们需要有更强的竞争力，他们需要一些其他方法。

这一道理也适用于工业领域。例如，亨利·福特二世（Henry Ford II）面临着去挽救一个走下坡路的企业时，他借鉴了其竞争对手通用汽车（Gerenal Motors）的管理概念。虽然有像福特（Ford）这样的实例，企业和公司仍不愿意在其自身范围外的领域中寻找更具竞争力的优势。由于害怕竞争，资源不足或保守，组织机构倾向于使用本行业中已经试用过和已经被证实可行的方法。这并不是说这些已经经过实践并得到证实的方法没有价值可言：米歇尔·史密斯（Michelle Smith）在没有使用田径训练方法之前就已经是一位世界级游泳健将了。

因此，寻找优势可被看作成功的三层金字塔。对运动员而言，他们首先必须竭尽所能做到最好。其次，他们要学习其他运动员的训练方法，把其他人所能做到的尽量做好。最后，他们需要从其他领域中学习经验，可以把生理学、心理学和营养学的技术运用到他们自己的训练计划中去。

这一道理也适用于公司管理。为了获得更有竞争力的优势条件，一个公司需要先审视一下自身。他需要检验一下自身的体制和工作方式，并进行必要的改进。还需要观察一下自己所在的行业，从其他企业中获得最佳方法，并努力自行实现那些最佳实践经验。最后还需要从其他行业中获取最佳实践，并试图实现它们。

观察其他的部门、公司或行业的过程，从根本上说是一种比较。如果不将收集的信息作为衡量自身业绩的标准使用，那么研究其他公司也就变得毫无意义。一旦进行比较，而且找出差距，其结果可以被用作制定目标的基础，以促进自身的提高。

这个建立在与他人比较基础上的成功金字塔就是目标管理的基础（图 2.1）。目标管理是公司中不同部门间的，或者在相同行业中的其他公司间的，或者与其他行业间在实践方面的比较。目标管理的目标是实现优势。

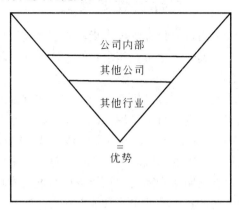

图 2.1　目标管理：成功金字塔

目标管理的定义

使目标管理与其他管理方法不同的是比较，尤其是指与外界环境的比较。但是，目标管理不仅仅只是简单的比较，在正式定义它之前，值得考虑一下它的基本组成成份。

它是有组织的

一些管理者会本能地将他们的部门或公司与其他人的做比较。有时这种比较也会不经意的发生，例如贸易杂志的发表和联盟表或统计表。然而这不是目标管理。目标管理是一种致力于实现优势的管理方法。因此，它必须是有计划、可实施和受跟踪的一种严谨并且有组织的方法。

它是持续的

公司所处于的商业环境可能变化很快。新的产品、过程和技术不断涌现，为了保持优势地位，公司必须对于这些变化采取相应措施。因此，目标管理是持续的，这样才能有效地实现其优势目的。

与最佳实践进行比较

一个公司如果仅仅将本公司简单地与其他公司比较，这并不一定有促进作用。因此，目标管理中重要的一点是与最佳实践进行比较。目标管理只有通过与最佳实践进行比较才能实现其优势。

它的目标是通过制定可达目标来实现整个组织的提高

通过与最佳实践相比，一个公司可以了解其目标管理的自身实践与最佳实践之间的差异，并不断缩小差距，最终实现优势。考虑以上这些基本因素，目标管理可被定义如下：

一个基于将公司的生产过程或产品与那些已被证明的最佳实践相比较的不断提高的过程，与最佳实践比较是用来制定最终获得公司优势的可实现的目标的一种方法。

目标管理发展史

对于大多数建筑业内人士而言，目标管理是指检测其他高度所用的基准水平。在管理方面，该词被更广泛的理解为建立一个可在相当长的一段时间里使用的衡量的标准。早在19世纪，泰勒(Taylor)就鼓励工作过程的比较。他的这种方法也曾和目标管理进行过比较。目标管理这一术语还被广泛的应用到计算机产业中，以阐明软件或硬件性能的级别。

上述之定义以及在此书中出现的目标管理一词，是一种全新的管理技术。它是由施乐(Xerox)公司在1979年创造的。施乐(Xerox)公司意识到在日本销售的复印机可以以同样的价格在美国制造。他们想找出其中的原因。他们购买了竞争对手的复印机，拆卸开来，分析其中的组成部件。此过程被证明是提高公司目标管理基准的成功的举措，施乐公司由此将目标管理的应用扩展到公司的各个商业单元和成本中心。起初，施乐公司中的非制造部门发现应用目标管理非常困难，因为他们仅仅处理商业过程，而没有产品可供拆卸和比较。但是他们逐渐认识到最终的产品是通过一定的过程发送出去的，而这些过程同样可与外界进行比较，以获得促进和提高。

尽管在施乐公司的应用取得了成功，目标管理还是有几年时间没有被多数人接受。有两件事促使这种局面发生了改变。第一件是罗伯特·坎普(Robert Camp)于1989年完成的著作。坎普(Camp)在施乐公司最初从事目标管理工作7年，并于1989年将其思想写进了他的著作《目标管理：寻找行业中促成优秀表现的最佳实践》。第二件事是1992年的Malcolm Baldrige国家质量奖，一个享有很高声望的奖，这个奖是专门授予那些引用了目标管理的竞争比较的美国公司(换言之，引用目标管理和竞争比较是美国公司参选的基本条件之一)。于是这两件事将目标管理引入了美国的公共主流之中。

目标管理的重要性在美国得以提高之后，此方法也开始在英国及欧洲其他国家引起广泛兴趣。例如在英国，贸易工业部发展了一套商业对商业交换计划，为来访者提供英国制造业和服务业中的最佳实践范例。只要支付一定的费用，公司可以访问最佳实践公司，并将最佳实践引入其组织内部。欧洲质量管理基金会(European Foundation for Quality Management)现在也已意识到了目标管理的重要性。

正因如此，建筑业对新的管理技术的推介往往比制造业在这方面落后。不过建筑业现在也已经开始研究和运用目标管理。到目前为止，最好的例子是建设信息技术研究(Construct IT study)和建筑研究学会(Building Research Establishment, BRE)所做的工作。建设信息技术研究(Construct IT study)所做的工作包括对11家主要建设公司在工地现场管理方面利用信息技术的程度与最佳实践相比较的目标管理。建筑研究学会(BRE)的工作则集中为建筑业提供一套目标管理方法论。

目标管理的类型

在本章的前面部分(图2.1)目标管理的成功金字塔中介绍，一个实业组织可在部门间，本行业其他公司，或者其他行业中寻找最佳实践进行比较。显然，这三种比较涉及到不同的程序和步骤，也会得出不同的优缺点。由于这个原因，它们通常被划分为三种截然不同的目标管理类型，每一种类型都将在下文中分别给予分析。在分析之前，需要先讨论目标管理中非常重要的一个概念——商业过程。

第二章 目标管理(Benchmarking)

一个实业组织可被分解为一系列职能部门。从商业角度来看，职能部门是指实业组织中具体部门的运作，例如市场推广，评估或者采购。所有职能部门都有自己的任务和产出。以评估职能为例，其产出可能是递交标书的总数。另一方面，商业过程是指职能部门内部所进行的一切活动。比如评估，其商业过程可能是投标决策，获得分包商的报价，或者是最终递交标书。这些过程又包含一些分过程。以最终递交标书为例，分过程可能包括对分包商报价的核对，不可预见费的估计，企业管理费用和利润的计算及提交标书。过程与职能间的不同在于过程是一种正在进行的状态，或者是将输入转变成产出的过程。所有过程的产出总量就是产品或者职能要达到的指标。各职能部门的产品总和即是最终产品，在建筑业中就是竣工的建筑物。

在上例中，评估部门的产出可被量性地看作是中标项目数量占总投标项目数量的百分比。然后，将这个结果与其他承包商的表现进行比较，看看是否有差距。在目标管理中，这种量性分析叫量化。问题在于，即使量化说明了表现存在的差距，也不能说明其中原因。相反，如果分析了商业运作过程，那么表现差距存在的原因就会很清楚。因此，多数目标管理方面的教科书都推荐要检验商业过程而不仅仅是计量或量化分析。

然而这又导致了一个来回重复的问题，那就是一个公司除非将有关内容进行量化比较，否则是无从知道表现差距的存在的。这就是目标管理要求思维模式转变的地方。由于超出本文范围以外的原因，如果过分强调量化管理这种手段来强调业绩，可能导致事倍功半。另一方面，如果商业运作过程及其分过程被看作是拼图中的组成部分，而且完整的图是最终产品，那么，所有生产过程的每一个过程的提高，或者至少是那些对于公司成功来说是最关键的过程的提高，将导致最终产品的改进。是过程造就了产品。因此，可以假设最佳过程将制造出最佳产品。

在制造业中，目标管理也许还是以最终产品或其成份的比较为基础。但是，在建设管理中及在本书中，目标管理一词被认为只与过程有关。以此来说，目标管理有三种类型。

内部目标管理

这是在同一公司机构中不同过程(部门)间的比较。

正如前面提到的，目标管理中重要的一点是在公司以外的环境寻找最佳实践。如果是这样，那么为什么公司机构要实施内部目标管理？答案包括四点。首先，在同一公司机构内部，各个商业过程可能是不一样的。原因也许是地域性的，也许是历史性的，也有可能是该公司被收购或合并。内部目标管理让公司认准自身的机构运作水平和公司内部的最佳实践，使得这最佳实践能在公司内部广泛运用。第二，内部目标管理提供了"外部"目标管理阶段所需要的数据。第三，内部目标管理，通过鼓励信息交流和新的思维方式，确保了目标管理可以被日后那些参与"外部"目标管理实践的人们所理解。最后，目标管理是以检验商业运作过程为基础的。因此，虽然同属公司内部，仍可以提供对该过程来说是外部的比较。

一个内部目标管理的范例就是在一个由主要工程部门、房屋住宅部门和粉刷更新部门组成的建设公司，比较各部门是如何处理设备的租用问题。

竞争型目标管理

这是同一行业不同公司之间商业运作过程的比较。这种目标管理有一个很大的优点就是它的适用性。最典型的就是对生产同类产品和相同的顾客群的两家公司的市场推广进行比较。但是竞争型目标管理的问题在于，由于我们只关心商业行为过程，而没有注意到竞争对手的最佳实践也许还不一定够好。例如，某一建设公司在设计和建造上有很好的声誉，然而这并不能说明他们的预算过程也比别人的好。如此的目标管理将不能使本公司达到优势。为了识别商业运作过程中的最佳实践，有时有必要走出本行业。

普通型目标管理

普通型目标管理仅考虑两家公司的商业运作过程的比较，而不考虑这两家公司是否属于同一行业。有一些商业运作过程是所有行业都有的，比如采购以及人事管理就是其中两个例子。普通型目标管理的优点在于它打破了思维障碍，并为改革提供了大量机会。它还拓宽了知识面，提出了许多创造性和启发性的想法。其缺点是难度大，时间消耗多，且花销大。

在分析目标管理的操作实施技术之前，需要考虑两个问题。首先是内部目标管理是否是竞争型目标管理的前提，以及内部目标管理和竞争型目标管理是否是普通型目标管理的前提。答案基本上是否定的。前面提到的目标管理的成功金字塔是一个理想状态；在开始研究其他行业之前，各机构都应全面了解他们自身的商业过程以及其本行业其他机构的商业运行过程。其实一个公司可以在不采用内部目标管理或者竞争型目标管理的情况下实施普通型目标管理，虽然，我们并不提倡这样做。另外，一个公司也可以只采用内部目标管理，而不用另外两种类型的目标管理。

第二点，也是值得提倡的一点，那就是，并不是所有的目标管理课本都使用相同的术语，这可能会导致一些混淆。表2.1列举了一些目标管理教材中所用到的不同术语。

事实上，用不同的术语描述基本相同的事物是无关紧要的。重要的是公司机构要选择最为适合的术语且一直使用同一术语。其选择可以与表2.1中总结的术语相同或不同。

表2.1 目标管理术语

作者	公司内部的比较	产品与产品之间的比较	同行业不同公司的比较	不同的行业间的比较
Camp	内部的	竞争型	功能型	普通型
Spendolini	内部的		竞争型	功能型（普通型）
Karlof & Ostblam	内部的		外部的	功能型
Blendell et al	内部的		竞争者及功能型	普通型
Codling5	内部的		外部或最佳的实践	外部或最佳的实践
Watson		反向工程	竞争型	过程
Peters	内部的	目标管理	目标管理	目标管理
本书	内部的	不适用	竞争型	普通型

注：Watson 也包括了另外两种类型：战略性目标管理和全球性目标管理。

图2.2说明随着目标管理的类型从内部型到普通型的转变，其难度、所用时间和成本以及创造力和机会都会增加。相反，当目标管理从普通型过渡到内部型时，成本、时间和难度会减少，相关性数据收集的难度，成果的可适用性和转换性也相应减小。

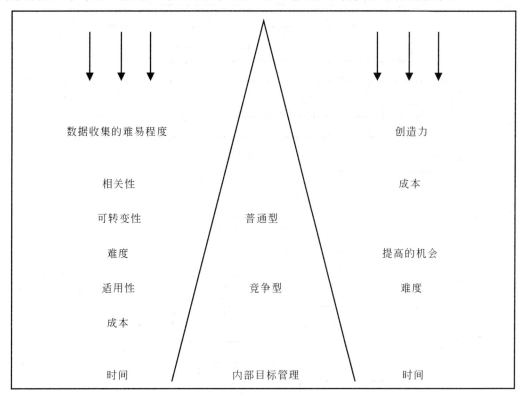

图 2.2　目标管理的类型

目标管理的过程

虽然在主要的涉及目标管理的课本中建议了许多种实施目标管理的方法，所有的方法都包含了相同的基本成份。本书建议一种九步骤目标管理方法，如图2.3。每个步骤将被分别予以讨论。

步骤一：决定使用目标管理

自由市场依赖于顾客可以选择供应商。这意味着当一个公司机构提供的商品和服务不受欢迎，或者未达到认可标准时，顾客不会购买这些商品和服务，这样一来公司很快会被市场淘汰。可是这种自由选择不存在于公司机构内部。内部环境的特征是商品或服务是由一个部门提供给另一个部门的，即使这些商品或服务不尽人意，也没有其他可能性做其他选择。这意味着虽然一个公司的业绩是盈利状态，且满足顾客的需求，但事实上公司机构的内部商业过程仍有发展提高的空间。改进这些内部过程将提高效率并提高最终产品的标准。因此目标管理为那些未暴露在市场压力下的过程提供了一个"安全网"。正像前面阐述过的，效率随着与最终产品的距离的增加而降低，这也是实施目标管理的主要原因。当然，还有其他原因，在下文中将进一步阐述。

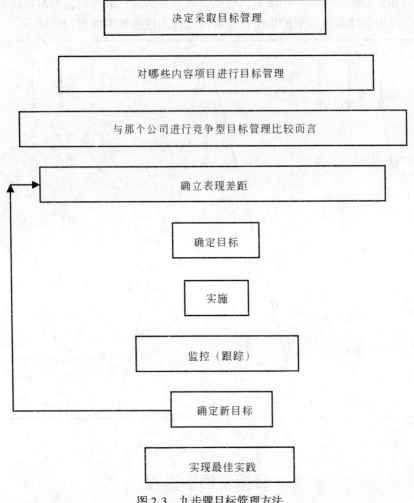

图 2.3 九步骤目标管理方法

变化的速度

如今的商业环境比以前变化得快。在建筑业，新方法、新产品和新顾客不断涌现。所有这一切都存在于外界环境中。一个不留意外界变化的公司将很快落伍，停滞不前，并慢慢地被淘汰出该行业。目标管理的一个基本要素是与外界环境相比较，而且，目标管理会迫使一个公司不断评估行业内的变化情况。

信息自由交流

在英国，其商业文化是这样的：信息不是自由交流的，并且有一种潜在假设，那就是信息是机密。这与其他国家的商业文化完全不同，比如美国，信息交换是自由的。"信息时代"的开始无疑要求英国对信息看法的改变，而且目标管理鼓励信息的自由交换。

以外界为中心和满足顾客需求

为了不被淘汰，任何公司的主要目标都应是满足顾客的需求。然而，顾客的需求和市场是存在于公司以外的。一个只注重内部的公司不可能了解顾客的需要，因此也不可能满足市场的需求。目标管理鼓励关注外界，以满足顾客的需要。

公认的过程的重要性

第二次世界大战之后，提供商品数量成为制造业的主要目标。在建筑业中，重点放在了提供大量住宅和工业建筑上。这种情况过去的 30 年里在逐步改变。日本的工业很快意识到了这一变化，并且开始生产比其竞争对手的商品质量更高的产品，因此日本的制造业基础得到了广泛的扩大。对日本制造业方法的研究显示，他们强调更多的是最终提供产品的生产过程。英国工业正在效仿此种做法，并采用了诸如全面质量管理，重组和目标管理等方法；所有这些方法都公认了这种通过检验过程从而改进最终产品的原理的重要性。

强调业绩表现的差距，阐述进行改变的必要

传统的商业评估方法试图在同一公司内部对不同部门职能进行比较。另一种业绩评估是基于对以往的或对预测业绩进行比较。这种评估只关注公司内部的。由于衡量的标准是公司目前最好的，或者是在过去实现的最好的业绩，这样的方法可使公司不去实现他们本来有能力达到的更好的业绩。对于公司业绩的真实衡量只可能通过与其竞争对手业绩的比较获得。目标管理通过在外界环境中进行的操作，表明了公司真实的差距。这样做强调了对寻求改变的需求和动机。

最佳实践揭示了如何实现变化

在目标管理中，计量这一术语被用来描述过程的可计量性。例如，对于同一个工程项目，两个建设公司标价的计算有不同的方法。从这些过程中得出的一种计量方法是中标率。如果将这两个中标率进行比较，业绩差异将突显出来。但是，这种孤立的信息有何用处？目标管理不鼓励将量化管理作为比较方法使用，因为它们只显示业绩差距，而且，如果一个承包商的运作过程已被确定为最佳实践，那么量化管理就变得无关紧要了。运作产生成果，这种情况下的结果是标价和投标书。选择尽可能好的运作会自动导致最佳产出。因此，没有必要专注于计量比较。致力于运作过程上的另一好处是它说明了差异存在的原因，而不像量化方法只阐明了表现差距。作为直接结果，它还指出了如何弥补差距。因此，对过程进行比较的目标管理不仅强调了问题所在，还提供了解决问题的方法。

最佳实践表明了什么是可实现的

对目标管理而言，一个明显的问题是为什么止步于最佳实践。与此相同，如果只以公司内部的最佳实践为基准，目标管理可能达不到实现其全部潜能的可能。相同的，以最佳实践为基准也是如此。没有依据说明，外部最佳实践就是本公司可实现的最佳实践。另外，存在着这样的争论，即如果一家公司效仿另一家公司的最佳实践，那么，这家公司是跟随者，而不是领路人。关于这一争论的答案包括两点。首先目标管理强调运作过程，在

一个公司内部可能有几百个过程和分过程。因此，公司可以只考虑过程，这些过程的结合将使公司获得商业优势。另外，以最佳实践为目标表明了什么是可实现的。如果一个公司定的目标高于最佳实践，那么将无法保证可以实现这个目标。

变化提供环境

正像前面提到的，目标管理要求商业文化转变。为了能够有效实施目标管理，公司需要改变对信息、内部商业运作过程以及竞争对手的看法。不过，问题在于多数公司从本质上讲不愿意改变。目标管理的提倡者认为目标管理的实施过程本身可以帮助克服这些障碍。因为目标管理是一个富于创造性的过程，它注重观察其他公司机构的商业过程，它就像促进变化过程的催化剂。因为目标管理清晰的显示了什么时候存在差异，而且它还可帮助激励员工进行必要的改变以缩小那些业绩差距。

目标管理可以发现其他行业的技术突破

这通常被认为是目标管理的一个优点，常引用的例子就是条形码技术。虽然最初这是零售业中的技术突破，但条形码技术已被广泛应用于图书馆、医院、保安和身份认证系统。

目标管理允许个人扩展其自身的背景和经验

从事某一行业的人通常自觉或不自觉地采用该行业的商业文化。虽然为了在该行业中能够有效的工作，这样做是需要的，但是商业文化同样可能会限制或抑制变化。与其他公司和行业合作可以发现，实施每一项任务都有多种方法，而且现有方法通常可加以改进。

目标管理关注目的

目标管理注重业绩差异，并制定缩小差距的目标。这些目标因此成为目标管理过程中参与者的焦点和目的。当所有职员都集中精力于这一目标时，此目标更易实现。

行业最佳是最可信赖的目标

在建筑业内，引用和实施制造业的成功管理技术的其中一个问题，就是，人们常说这两个行业是不一样的，不管该管理技术在制造业内有多成功，也不能直接搬过来。但是，不管什么行业，都有商业过程，集中于这些过程中的最佳实践可以提供一个可以实现的目标。这同样适用建筑业。

步骤二：对什么内容实行目标管理？

这是目标管理中最难的部分。也许正是因为这个原因，很多关于目标管理的教科书都没有谈及这个问题。本书着重强调了目标管理中最为基本的一环就是商业运作过程，但并非所有人都同意此观点。例如皮特史（Peters）定义了目标管理的三个类型：策略型、操作型和统计型。策略型目标管理针对的是文化、人、技巧和策略（帕斯托（Pastore）[14]也认为目标管理可以在策略层面上开展。）操作型目标管理针对的是目标管理的方法、步骤和商业运作。（这也被称为过程目标管理。）最后是统计型目标管理，它是对于公司业绩进行数字或者统计上的比较。

认为可进行目标管理的对象不只是商业运作过程的这种思想只会使本来简单的问题复杂化。"最佳企业文化"是不能进行比较的。即使可以，这样做也没什么意义，因为文化是不可能轻易被改变得了的。文化，包括存在于公司内部的亚文化，是一种无形资产。另外，一个公司生产的产品类型与其企业文化之间存在着密切联系。因此当新提供的产品仍然有所不同时，实施文化方面的目标管理的意义不大。将目标管理划出一类进行纯统计的或数字的比较同样会将问题变得复杂。正如前文解释过的，业绩的数量化是过程的结果。这些数量间的比较除了说明了业绩差距的存在以外，并没说明其他什么。因为目标管理被定义为商业运作过程的提高，这种统计性的比较本身不能称作目标管理。

当然目标管理的方法很可能会继续发展。将来的某一天，也许可以对公司策略的某些方面甚至对人进行目标管理。但是，在本书中，作者相信，目前公认的目标管理方法只对产品及产出运作过程进行比较。在建设施工管理这一更窄的范围里，目标管理一词仅包括对商业运作过程的比较，因为最终产品(即建筑物)，由于其多样化且太复杂，而不可能进行有意义的比较。

那么如何定义商业运作过程？

任何商业行为都可以被划分为三个阶段：即投入、过程和产出。这些产出的有机结合导致了最终产品。为了便于组织和管理，可以根据各阶段的职能将其分组。从而获得特定产出成果。图 2.4 和 2.5 说明了建筑承包公司是如何运用这概念的。图 2.4 说明了公司在生产过程中如何划分职能部门。图 2.5 说明了分包商管理职能部门的详细运作过程和分过程。

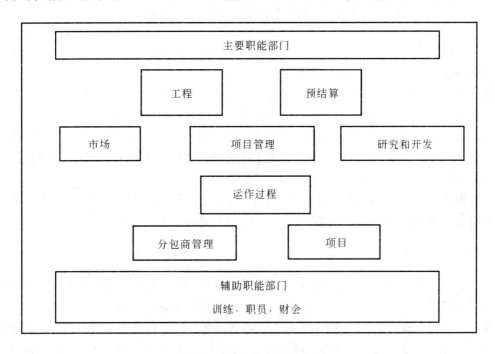

图 2.4 承包公司的职能部门

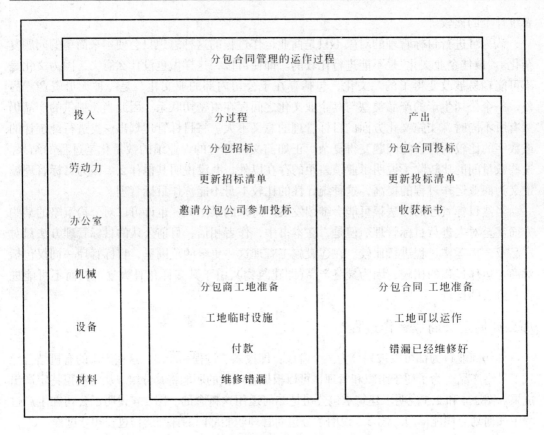

图 2.5 分包合同主要功能中所包含的过程

在图 2.5 中项目管理是一种商业运作功能,而分包合同管理是其中的一个过程。此过程中还包括了分包合同招标和工地准备工作等分过程。在招标这一分过程中,又进一步包括了保留最新的招标单位清单和收邀请分包公司参加投标的分过程。这些分过程有定量的资源投入,并且有一定产出。在这个例子里,其成果就是完整的招标清单和从分包商那里获得的标书。所有这些产出的总和是运作过程的结果,而过程结果的总和又是功能产出。功能产出的总和就是最终产品,即竣工建筑物。这是目标管理的基础:通过改进过程来改进最终产品。

然而随之产生的问题是一个公司可能会有几百个甚至上千个分过程,该对哪一个过程进行目标管理呢?可以采用不同的方法,这些方法将在下文中加以阐述。

确定商业功能的产品

正如前文提到的,一个公司的职能部门或行政部门就是他的产业职能部门,即这正是该对什么进行目标管理的良好始发点。因为商业功能比商业过程更易识别。例如,承包商 X 在项目管理方面很优秀。而另一个承包商意识到他自己的项目管理不够好,于是可能采用商业功能作为目标管理的开端。对部门功能内的运作过程进行调查可能会使其发现,他的问题在于分包商的管理上。因此,他也许对分包商的管理过程进行详细分析,并将其与更为成功的承包商的做法和过程进行比较。因此,识别更好的产品功能,就会导致更好的生产出该产品的运作过程。

此方法的问题在于，在功能和过程之间没有明确的界限。在上例中，分包合同管理被称为过程，但是它同样可被描述为功能。虽然过程和功能的划分通常是明确的，但仍存在无明确特征区别的灰色地带。在这样的情况下，唯一的答案就是依靠对公司、行业和管理的经验。

关键的成功要素

在任何公司内部都有关键性的成功要素。这就是那些如果不加以有效操作，将损害公司的运作的因素。就房屋建造者来说，市场部门就是关键成功因素，而对总承包公司的土木工程部而言，关键成功因素是按时完工。查证这些因素是实施目标管理的一个好的始发点。除关键成功因素外，目标管理新选择的内容必须是那些能对公司的业绩的提高产生巨大影响的。例如承包商也许知道营业额是成功的关键因素，但如果该公司已经在其最大可能的营业额情况下经营，那么，就算继续在此方面努力，也不会带来更大的收益。关键性的成功因素可被比作帕累托(Pareto)规则，即80%的成果来源于20%的行为。对目标管理来说，关键成功因素即指那20%的行为。让人吃惊的是，没有几个建筑工程管理人员真正知道他们自己公司中的关键成功因素。因此当其公司遇到困难时，经理们会立刻削减开支，因为他们认为这是唯一能提高业绩的途径。他们相信削减成本并维持原产出，价值便会自动增加。事实上，这样做除降低了士气及服务质量以外，没有其他收获。

施乐公司的问题

为帮助认清哪些地方宜采用目标管理来检验，坎普(Camp)推荐了一系列问题，其中的大部分被收录如下：

哪些因素对于你的商业成功至关重要

　　对承包商来说，关键成功因素也许是：
- 营业额
- 中标率
- 路桥等基础项目与楼宇工程项目的比率
- 房屋售出的数量

哪些部分最易出现问题
　　其答案也许是：
- 对分包商的管理
- 按期完工

提供哪些产品或服务
　　最典型的是：
- 房屋

- 维修服务
- 小型楼宇项目

哪些因素会使顾客感到满意

如果对建筑业客户进行调查，会发现以下因素关系到其满意程度：
- 按期竣工
- 最低造价
- 保证造价
- 质量
- 职业健康和安全
- 与环境和谐的发展
- 建筑美感
- 生命周期的成本
- 耐久性

什么是竞争的压力？

这些可能是地域性的、国家性的或者国际性的，压力可能来源于想跻身于新市场的小型建造商，或者是大承包商在市场不景气时寻找小型工程项目。

哪些是主要成本/什么职能部门在成本中占百分比最高？

这可能是总部，永久性职员，公司附属的加工厂或者市场营销。

什么方面最需要改进？

这可能是职员的培训或者材料订购。

哪些部门最有影响，或最有潜力，可以使其在市场竞争中脱颖而出？

这通常是最难回答的问题，因为这正是确定一个公司超出另一个公司的优势之所在。

询问顾客

另一种决定哪些地方宜采用目标管理的途径就是询问顾客。在前面的例子中提到，一个公司也许将项目管理视作具有发展潜能的领域。但是谁是这一方面的顾客？是业主、现场监理或者是分包商？答案很可能是后二者。讨论这些将会弄清楚管理上的问题在哪儿，这些问题存在的地方很可能成为需要进行目标管理的地方。顾客并不一定是指最终使用者。在目标管理的例子中，顾客是指该过程的顾客，也就是整条链中的下一个人，以及获得了该过程的成果的人。

从上文可以看出，对什么过程需要实施目标管理这一问题没有确定的答案。以上提到的任何方法或者是方法的组合都可用来寻找答案。像本书中提到的许多方法一样，目标管理是一种软性系统。因此答案将是因公司而异，而且联结于该公司的企业文化。唯一真实的答案来自于对于公司的全面了解。

第二章 目标管理(Benchmarking)

在决定对什么方面实施目标管理时，需要进一步考虑的问题是，具体到什么程度。如上文指出，一个商业过程可以被分解为分过程，即这些分过程还可以被继续划分。选择的详细程度完全取决于各个公司，并且它将随着公司的规模和可供目标管理支配的资源的多少而变化。

步骤三：与哪些公司进行比较？

此问题的答案既简单但又非常困难。简单的地方是指比较对象应该是那些商业运作过程最佳的公司。困难在于如何找出谁是这样的公司。对公司的某一些职能，诸如市场营销，很直接就可以找出来。谁是最佳企业，具体对房屋建造业而言，可以很容易搜集销售信息，广告和宣传册。另外可以参观工地现场，和检查样板房以及现场营销情况。如果可以获得相关资料，就可对房屋购买者进行调查，并且可雇用市场咨询顾问公司分析调查结果。一旦信息收集齐全，就可以做出最佳实践企业的评判。

在总承包商的土木工程部门，确定最佳实践企业的过程变得越发困难。因为产品不是被公众购买的。他们完成的项目数量有限，没有宣传册，关于完工项目的数据的公布也很少。那么如何获得用于评价最佳实践的信息呢？

寻找在具体的部门或过程中领先的公司的唯一方法是收集数据。多数关于目标管理的课本都列出了确定最佳实践公司的信息源，包括：

- 特殊奖项和引用
- 媒体关注
- 专业协会
- 独立报告
- 口碑
- 咨询顾问
- 目标管理网络
- 内部信息
- 公司资源
- 图书馆
- 专家及研究
- 通过邮件和电话方式进行问卷调查
- 直接现场调研
- 核心工作组
- 特殊兴趣小组
- 职员、消费者和供应商
- 外来数据资源
- 学术研究所
- 投资分析
- 互联网
- 网上数据库
- 期刊杂志

步骤四：确立业绩表现差异

一旦确定最佳实践，应该主动与拥有最佳实践的公司联系，并使之成为目标管理的合作者。对于大多数的建设管理专业人员来说，这是一大难题。因为都不情愿与其他公司或竞争对手接触。同样的，最佳实践企业也不情愿提供有助于提高竞争对手的信息。正如史本多林尼（Spendolini）所概括的，大多数私人企业的雇员，视击败竞争对手为己任，而不是培训对手。

这种态度是可以理解的；毕竟它是基于多年的企业实践基础上积累出来的。但是通读本书的所有章节，到现在为止，有一个问题应该很清楚，那就是所有已阐释过的技术，不管是全面质量管理还是价值管理，都要求在建筑业的运作方面有一定程度的文化改变。在过去的十五年以来所有行业的市场都在不知不觉中发生了很大的变化。现在的市场是全球性的且将注意力集中在顾客身上。建筑业较慢意识到这一点，但是外部竞争意味着再也不能忽视这种改变了。如果企业要生存，就必须改变。改变的其中一部分内容无疑要求公司间在一定程度上的合作。

对于一个公司如何成为一个目标管理的合作伙伴的具体要求并没有一定规定。只取决于公司的集体情况。关键因素在于这是一个合伙人制。双方都希望从中有所收获，哪怕这种经验只是如何实施目标管理。

确定了最佳实践企业并且达成了目标管理合伙协议，下一步就是对研究过程做全面记录。方法有许多种，诸如流程图以及过程图表。本书由于篇幅有限，对此不进行详细的介绍，但是所有优秀的管理学教材通常会仔细讲解这些方法。

在本章的前面并不赞成使用量化手段或者将过程的成果量化的方式。然而在完整记录了检验过程之后，可以使用量化手段。在一些实例中，量化手段是至关重要的。诸如中标率，按时竣工的项目数量或者分包商的付款要求。但是在一些情况下，也许简要的记录过程并将研究对象的过程进行比较就已经足够了。公司内部的投标记录就是一个很好的例子。量化的使用可以有好几种形式：比例，单项合同的成本，成本占营业额的百分比。不管采用哪种方法，公司都将在就自己公司的一些具体过程的业绩与最佳实践公司的相应业绩比较之后，得出如下三种结论中的一种：

平等：两个公司有相同业绩表现。
正业绩差距：本公司的业绩表现事实上优于最佳实践公司的业绩表现。
负业绩差距：本公司的业绩低于最佳实践水平。

步骤五：设定目标

最可能出现的结果将是该公司的业绩表现低于作为目标管理对象的公司的业绩表现，存在负业绩差距。此种情况下，该公司需要想出办法弥补差距，除非此差距很小且可以一步解决。旨在通过一系列的步骤实现最佳业绩表现的，有必要设定一个目标。

本文不打算详细解释公司内部目标设定的步骤，因为在任何一本优秀的管理教材中都可以找到这些内容。总之，目标必须是经过计划的，实在的，而且可实现的。这在目标设定过程中，必须要与员工进行交流且要求他们的参与。

步骤六、七、八：实施-监督-设定新目标

目标管理的下一阶段要求实现第一目标的计划得以实施和监督，然后需要设定新的目标。在这里不再叙述这些步骤，因为从多数管理教材中可以找到相关内容。另外，多数公司都有其自己实施计划和监督进度过程的步骤。这里虽然没有包括这些内容，但并不说明目标管理中这些步骤不重要。它们对于成功至关重要。有计划不实施比没有计划更糟。

步骤九：实现最佳实践

实现了最佳实践意味着公司已达到了与目标管理的公司同等的水平。然而目标管理过程并未就此结束。正像前面提到过的，目标管理是一个连续的过程。像全面质量管理一样，它是一个公司所拥有的，而不是公司偶尔为之的事情。为了获得优势，公司需要在其所有主要过程中实现最佳实践。只有通过不断的寻找进一步的改进措施才能达到此目的。

本章第一部分讲述了目标管理的步骤。第二部分则讨论与实施有关的内容。

目标管理工作组

目标管理，与价值管理不同，不一定是团组活动；它可以由个人独立完成。但是考虑到分担工作量，且有助于对目标管理的认可和交流仍建议设立工作组。有几种不同形式的目标管理工作组。

工作小组

这些小组已经存在，并且包括了现有部门或小组的成员。

部门间小组

这些小组的形成是将所需的专业人才协调到一起。这些小组通常在目标管理研究结束后解散。

临时工作组

这些小组更为灵活，且由任何有兴趣加入目标管理工作的人组成。小组成员来去自由。除非像施乐公司一样是一个成熟的目标管理公司，这种目标管理工作组不会有良好的工作表现。

目标管理行为准则

美国产品质量中心属下的国际目标管理所和目标管理战略规划研究委员会已经制订了一套目标管理行为准则。其中包括如下九条基本原则：

合法化

这条原则排斥任何与贸易法相违背的内容，例如用欺骗等手段操纵投标。

互　换

实行目标管理者不应向其他公司索取他们不愿向人透露的信息。

机密性

从目标管理合作中获得的有关目标管理合作者的任何信息都不应向他人泄露。

用　途

从目标管理研究中获得的信息不应用于过程改进以外的其他任何目的。广告宣传就是最明显的例子。

第一联系方

目标管理实施过程中的第一联系方应为负责目标管理的人，而非商业过程的主管。

第三联系方

除非预先取得允许，不应与其他公司分享对方公司的联系人和详细联系方式。

准　备

直到所有的准备工作已经就绪，才可以联系目标管理的合作对方。

结　束

除非可以进一步跟进，目标管理合作者之间在目标管理研究结束后，不应有任何承诺。

理解与行动

在开始研究目标管理之前，合作双方均应了解目标管理的所有过程，且清楚说明各自的意图，特别是关于信息的利用。

法律条文的规定

以下是目标管理实施过程中主要的法律规定：

- 罗马条约第三篇第一章，特别是第 85 款（含 1,2,3 条子项）和第 17 条规则（第 4 款）。这些协议适用于所有欧盟国家，且协议中说明关于价格的确定、分配，以及因不利的竞争条件或地域限制、市场分离等原因对第三方予以不公平待遇等做法均是被禁止的。

本国法律包括：
- 《限制贸易实行条例 1976》
- 《公平交易条例 1973》
- 《竞争条例 1980》

目标管理：主要议题

本书不想"推销"目标管理的方法，而是希望就建筑业中如何应用这些方法给予客观的评价。因此，下文讨论的是现阶段的思想，这些思想也许会引起那些从学术角度思考目标管理的人们的更大的兴趣。

目标管理和全面质量管理间的联系

绝大多数关于目标管理的文献都认为目标管理与全面质量管理一同实施时，将发挥最佳功效。杰克逊（Jackson et al）等人曾提出如果没有全面质量控制方案，目标管理的实施只能是时间与金钱的浪费。这种说法真实与否，尤其在建筑业领域，还有待证明。

目标管理中的革新部分

有人批评目标管理并不是一项创新方法，因为它是跟随而不是领先。然而，沃特森（Watson）认为目标管理只是综合体中的一部分，且优势来源于质量、技术和成本：即质量超越竞争，技术先于竞争，成本低于竞争。本文作者同意此观点。目标管理并不试图成为革新的工具，它只是改进过程的工具。与此相反，本得尔（Bendell et al）等认为目标管理技术可以是创新的，并且目标管理的第四种类型，即消费群比较目标管理的内容，超出了消费者希望或要求的范围。从这个角度讲，它不仅仅是一个改进过程，也是一个革新的过程。

使用量化法的问题

在《世界废物》（World Wastes）杂志上介绍了美国的几个主要城市的废物回收服务和机构，使用了量化制。这是一个说明为什么进行目标管理时要避免使用量化管理的典型例子。论文描述如何用目标管理来"将一个服务传输系统与其他系统比较"，并且将垃圾收集的数量与废物回收服务的成本进行比较。但是由于此过程既未选择最佳实践，又未针对过程进行改进，因此它不能被称之为目标管理。它仅仅是一种简单的数量上的比较。

量化法应该在将企业的生产过程详细记载之后再使用，但在现实中，这样做颇具难度。管理者们希望在他们调配资源来弥补业绩差距之前能够估计出差距的大小。对文化转变需求的争论不会对只掌握有限资源的管理者产生太大影响。在这里，答案就是如果没有其它方法，量化法可以在目标管理的前期阶段使用。但是在使用量化法时需要多加留意，并且需要在目标管理归档记录过程中重新计算。

商业合作

对于多数管理者而言，这是目标管理中最困难的地方。关于商业合作，有两个基本争论点。首先是建筑业的目标管理的核心存在于运作过程中，而运作过程不像制造业产品那么灵敏。第二点，也是本书一再强调的主题，就在建筑业中非常需要文化转变。按照沃特森（Watson）的观点，目标管理是不断进步的协调的全民挑战。许多公司已经意识到未来的成功是依赖于全球性市场和国家生产力的提高。没有合作就没有目标管理。任何一家公司，如果不能接受此概念，就无法实施目标管理。

该走多远？

在最佳实践研究中，一家公司，该走多远（做到什么程度）才能达到目标？格林尔(Grinyer)和史密斯(Smith)建议研究应是全球性的。然而，这是不现实的。对于一个中型建筑承包商而言，通过研究整个世界来寻找改进手段的做法是不可行的。没有几家公司有进行如此广泛和详尽的研究的能力。目标管理合伙人必须是具备相应规模和地位的对象公司，且能提供高度适用性和相关性。

日本 Kaizen

一些文章，常把目标管理与日本的 Kaizen 进行比较。但这种比较并不正确，因为 Kaizen 是一个更广义的概念，它不仅仅依赖于与外界比较的系统。本书的全面质量管理一章，对 Kaizen 进行了更为详尽的阐述。

目标管理的缺点

兰库斯(Lenkus)研究了目标管理的缺点，并指出目标管理是复杂、低效、费时且昂贵。他还发现在目标管理合作者的可靠性上也存在问题。虽然他能看到目标管理量化的优点，但他认为，目标管理的实施过程是困难的。同时他还认为，经理们永远都希望在对改良过程投入资源之前，先了解具体的量化的差距。其他缺点还包括占用的时间长，以及目标管理研究的进展也许跟不上技术更新的速度。

当前的研究

目标管理，特别对建筑业而言，是较新的技术。因此对此项技术的研究也还不多。下文仅对目前已取得的研究成果进行讨论。

有人已经对目标管理工作在建设项目的计划前期阶段的实施进行了研究。这项研究涉及了62个建设项目的计划前期阶段中的四个主要分过程，包括组织、挑选项目、制定项目定义包括以及决定是否将项目进行下去。该研究工作收集了这些过程的数据并对之加以分析。此研究工作存在一些有争议的问题。首先是对这62个建设项目的过程和次过程的检验是基于理论模型基础上的。由于模型是纯理论性的，因此不能对与之相关的建设项目的表现加以评价，因为如果模型是不正确的，那么由此得出的所有评价也会不正确。研究中产生的第二个问题是建设项目及其过程是否可以采用目标管理。涉及到的设计和承包公司的数目使过程变得非常复杂。一个建设项目管理小组是建立在为该项目服务的一次性的过程的基础上的临时性的小组。因为设计和施工的时间有限，加之许多组织都加入了这个过程，那么怎么可能建立最佳实践呢？就算一个项目的过程被确定为最佳实践，那么对于新的业主，设计咨询公司或者承包商而言，也肯定不可能完全重复最佳实践的做法。这引起了一个非常严重的问题，那就是目标管理在建设项目管理的应用的未来研究的问题。具体来说就是建设项目能否采用目标管理，或者是否只能对存在于项目范围以内的个体的组织实施目标管理。

目标管理研究中的另一个有趣的方面是目标管理所赖以生存的技术和文化之间的关系。沃特森(Watson)认识到因为目标管理的未来是实现全球性应用，那么就应该将目标管理放在国际贸易和文化层面上。沃特森认为未来目标管理的需要是在那些文化差异之间建起桥梁。然而不仅仅只有国家间的文化差异可以影响目标管理,企业文化同样与之有关联。企业有三种运作行为:基础型、革新型和竞争型。在革新的气氛中,一个公司选择了冒险并希望消费者会喜欢。索尼随身听(耳机)就是此类型的范例。对于竞争型行为而言,竞争是基于对多种产品性能的直接比较基础上的。与革新型行为不同,它对于消费者的要求不是猜想的,而是已经有所了解。基础型行为满足的是消费者的最基本期望而且所供产品只包含基本特征。毫无疑问的,这些企业文化存在于制造业当中,但是它们是否也同样存在于建筑业中呢?如果不会,那么建设公司实施本行业外的目标管理有价值吗?由企业文化带来的障碍会不会变得过大呢?

史本多林尼(Spendolini)对于文化差异对目标管理造成的影响做了一个精辟的比喻。这就是他所称之为超出盒子范围的思考。公司准备走的路线超出该盒子越远,他们所遇到的文化差异就会越大,但是回报也可能越大。虽然没有真实的证据证明其正确性,或者可实施性,但这是一个有趣的想法。

前面已提及建设信息技术中心(Construct IT Centre)在目标管理方面的研究工作,该项目将11家建设公司在工地现场使用信息技术的情况进行比较。这些公司从信息技术使用的角度,将其最佳建设项目推荐出来。此项研究选用一家工程公司作为标竿(比较对象),因为其商业过程与建设公司的颇为相似。该项研究的报告就有效的建设工地运作过程的管理提出了一些建议,这些建议包括提高信息技术重要性的策略,建设公司对其商业运作过程的重新检验,对信息技术进行更多投资,以及增加使用信息技术的培训。

案例分析

在其著作中,格兰·彼得(Glen Peters)对消费者服务领域所采用的目标管理进行了检验。顾客服务在建筑业中一直是一个薄弱环节。作为一个例子,他对目标管理做了一个简单试验,他将每一家普通的房屋建造商的顾客服务的方法与汽车制造商采用的方法相比较。之所以选择汽车制造商作为比较对象是因为其在汽车的市场营销,产品的特点,以及所提供的客户服务等各个方面都明显的优于房屋建造商,而且双方经营的都是昂贵的,而且顾客不会经常购买的商品的生意。正像前面解释的,由于目标管理关注商业过程,而过程打破了企业障碍,所以将房屋建造商的客户服务与汽车制造商的客户服务进行比较是有效可行的。我们在得到格兰·彼得本人同意的情况下,将他的调查表中的问题和调查结果从她的著作中查选出来,汇总于表2.2。

很明显,从如此小规模的调查中是无法得出确定结论的,但是,正像预测的一样,汽车制造商比房屋建造商,更多的关注消费者。他们不仅注意到得到和失去消费者所付出的代价,他们还拥有单独的投诉部门来记录每一次投诉。此外,与房屋建造商不同的是,他们的所有前线职工都接受过关怀消费者训练。除此之外,更有趣的是,当一家预结算公司也被邀请回答同一份问卷时,而他们却交回了未完成的调查表,解释说他们不理解这些问题。

表 2.2　建筑业和汽车业，在顾客服务方面的比较

问题	汽车公司	房屋制造商
你如何鉴别谁是你的顾客	数据库 工作小组 量化书面研究报告 访谈	数据库 目前使用的账号 计划
你如何获得潜在顾客的信息	直接市场推广 插入 通过经销商	书面报告 电话调查
你多久检查一次你所有的关于潜在顾客的信息	一年	6~12个月
你能预先确定的顾客群占总的顾客群的百分比是多少	不知道	70%
你用来明确顾客群的调查开支占总的市场研究预算的百分之几	100%	10%
记录在案的投诉占所有投诉的比例是多少	100%	40%
你如何保存投诉信息	公司范围的计算机系统	每个工地的手记
你是否计算失去一个顾客所带来的损失	是	否
你是否计算赢得一个顾客所需要的成本	是	否
你是否证实了产品投诉最多的部分	是	否
有百分之几的前线职员接受过培训	100%	80%
你如何分析投诉	投诉处理部门	请前线职员提出改进方法
你是否会向顾客解释他们的投诉所带来的结果	是	是或否
你是否会对投诉顾客做出一定的补偿	是	通常否

结　论

作者在撰写本章时发现的一个非常有趣的内容就是约诗莫力(Yoshimori)的论文。在论文中，他研究了不同国家的高级管理者如何看待他们公司的所属性问题。在英国，70.5%的高级管理者将干系人作为他们优先考虑的对象。在日本，只有2.9%的认同。然而在回答是否要对干系人，其中包括雇员给予优先考虑这一问题时，29.5%的英国管理者采取肯定态度，而日本，这样的管理者的比例则占97.1%。

坎普(Camp)的书，研究目标管理的第一本教材，讲述了施乐(Xerox)公司的目标管理历史发展。书中他详细地说明了施乐(Xerox)公司计划的总目标是通过获得质量来取得领先地位，而这是通过三个部分加以实现：即目标管理，员工参与和质量过程。然而坎普(Camp)之后的教材大多没有提到这一点，而将目标管理集中作为一项独立技术。这一点在杰克逊等(Jackson et al)的论文中得到了最有力的强调，该论文对现行的许多目标管理教材做出了评价。该论文还为某些主要的目标管理论题的全面性打分，其中包括全面质量管理过程。但是没有包括员工参与这一主题，因为目标管理中的这一方面的重要性已被削弱。许多目标管理教材力荐日本做法来表示目标管理可取得的收获。虽然这是事实，但另

一种事实也许是日本的成功部分源于公司内部员工参与的原理。目标管理不能忽略这一部分。甚至在目标管理的发源地，施乐公司，员工参与也被看作是获得商业成功所采取的三部分计划中的核心部分之一。然而这一点在目标管理的发展过程中似乎被遗忘了。如果它继续被忽略下去，那么，在作者看来，目标管理将会衰竭，并会加入到被缩略的行列中，而此现象曾在商界中发生过。

第三章

重组(Reengineering)

导 言

尽管重组是一门新的学科，对某些人来说，重组的含义已经有些模棱两可。起初这个词有两种拼写法。有些人喜欢采用带连字符的拼法"重-组"(re-engineering)，有些人采用无连字符的写法"重组"(reengineering)。被誉为重组的引导者的翰默(Hammer)倾向于用无连字符的写法。我们在本书中也采用此写法。在大部分常用的词典中，例如韦氏(Webster)词典，这两种写法都没有包括在内。

另一个不明确的地方是，究竟该用重组一词，还是商业过程重组一词来作为广义的重组行为。因为大多数有关参考文献是以商业为背景的，商业过程重组最为常用。对此我们和翰默(Hammer)立场是一样的，倾向于把重组当作一个广义的重组行为。因为建设过程重组这一术语是近年来才创造的，而且本书也想让读者区分商业过程重组和建设过程重组。

目前，有很多文章都强调重组的作用。在翰默(Hammer)和珊比(Champy)所著的教科书《公司重组》的扉页中，他们提出了"在重组公司的过程中，翰默(Hammer)和珊比(Champy)为现代企业做的事情和亚当·史密斯(Adam Smith)在两百年前为工业革命编著的《国家财富》(Wealth of Nations)是做同一样的事情——翰默(Hammer)和珊比(Champy)通过对工作本质的研究，创造商业竞争转向的惟一而且最好的希望"。

本书自始至终尽可能做到中立和客观。但是，在本书中涉及的所有概念中，重组是最难让我们保持中立的。主要原因是许多重组的倡导者认为重组是一个革新而不是进化的概念。因此，多数描述重组的语言都带有一种感情色彩。例如，翰默(Hammer)常常引用的一句俗语"不要任其自生自灭，而要消灭他"是增强重组的革新性说法的典型。许多关于此课题的书本称之为诸如"激进性变革，引人注目的效果"之类的话语。翰默(Hammer)和珊比(Champy)用了上述两种表达方式定义重组，"为了显著提高工作成果，而进行的商业过程的彻底的再设计。"莫利斯(Morris)和布兰登(Brandon)很极端的认为"你可以选择重组，也可以选择破产。"在像莫利斯(Morris)和布兰登(Brandon)等人激进的评论下，我们很难对重组做出冷静的，客观的解释。但是 COBRA (Constraints and Opportunities in Business Restructuring——an Analysis——关于商业再结构的局限与机遇的分析报告)对这种激进的言辞提出了对立的观点。这是欧洲委员会提出的一个议案。COBRA 的领导与协调人库森—托马斯(Coulson—Thomas)教授，描述此工程的背景："创造力与想象力是非常宝贵的。根据它的理论和措辞，商业过程重组是关于改变步骤，而不是逐步提高，是革新而不是进化。然而当商业过程重组被用于改进现状，例如降低成本，缩短生产周期，压榨工人更多的劳动时，它为激进的变革和创新又做了什么呢？"

第三章 重组(Reengineering)

我们想为读者做的，是在介绍重组的概况和探讨重组在建筑业中的用途时，尽可能地保持中立。

应该承认，我们更倾向库森—托马斯的观点，也就是重组不是取得根本和基本改变的惟一方法，而且创造性思维，目标管理，文化改变以及革新可以脱离重组而独立进行。但是重组还是一个很重要的管理概念，它曾经使大型商业集团的业绩取得了相当显著的提高，它的优点值得认真地思考。

重组的起源

尽管1990年翰默(Hammer)的在《哈佛商业回顾》(Harvard Business Review)杂志中的一篇文章"重组工作：不要任其自生自灭，而要消灭他"通常被人们认为是重组的起源，实际上，重组的起源可以追溯到20世纪40年代，英国特威斯托克(Tavistock)学院将"社会技术系统方法"(Social Technical System Analysis)用于英国的煤矿工业的实践。社会技术系统方法的本质是，技术和社会系统应相互协调来实现最优整体系统。20世纪80年代社会技术系统方法引入管理费价值分析(Overhead Value Analysis)，这种分析强调把重点放在完成的工作量上，而不是放在做该项工作的人身上。瑞迪(Rigby)认为，管理费价值分析就是重组的前身。

社会技术系统方法，管理费价值分析和重组方法的基本主旨就是工作过程需要重新进行设计。翰默认为，进行重组并不意味着我们最开始设计的运作过程效率低，而是因为这些商业过程由于时间的推移和技术的发展而不能与时俱进。

"随着时间的推移，每家公司都形成了自己完整的工作程序，却没有人回顾全局。今天，如果大多数公司都可以从头开始，他们将为自己创造一个与现在完全不同的方法。"重组赋予社会技术系统方法和管理费价值分析以新的含义，是突变发展的概念。重组在于获得一个飞跃，而不是细小的改变的进步。

建筑业中的重组

在建筑业中考虑重组时，首先要确定建筑业是不是一个特殊行业，对待它是否应该有别于其它商业。对此问题，既有赞成意见也有反对意见。当然建筑业有自己的特性，并且其产品也有独特性。但是此产品的独特性并不足以作为行业特别的论据。汽车业与家用电器业各自生产有明显区别的产品，一个生产必须在室外操作的汽车，另一个生产在室内使用的家用电器，但是没人会否认两个都是商业过程。

经常有这样一种观点：建筑业是一个高度分隔的行业，高度分散，受限于资本，操作在一个个不同的项目的基础上进行等等。我们越是研究建筑业与其他行业的区别，就越会确定建筑业需要重组。对建筑业来说，最好的方法就是把自己当作国际商业团体中的一员，与麦当劳、施乐、丰田一样以同样的基础起步。虽然，总体来说建筑业的规模比其他跨国集团要小，但是在西方经济中，它的产值通常要占到国民生产总值的8%到10%。这就意味着，10%的建筑业绩的提高就会使国民生产总值增长2.5%。

对于重组，我们希望读者能将建筑业视作是商业社会的一部分，而不是独立于其外的。这并不是说建筑业的每件事都得效仿其他行业，而对整个商业社会毫无贡献。

现在，商业过程重组应用的趋势是集中于企业内部："商业过程重组的应用成了排他性的内部操作,如果有必要,公司自己应该迎合消费者的要求进行决策,技术投资和资源组织,而且要最小限度地受外界因素的影响"。至于重组是否能够在整个建筑业界中应用,而不仅限于单个企业,是一个需要强调的挑战。"一个行业,例如建筑业,就是对商业过程重组的一个主要挑战。这主要是因为控制此行业的复杂的商业关系,再加上外部因素在建筑企业如何开展生意中所起的关键作用。"

我们的观点是,尽管建筑业有它独特之处,它仍是商业社会的一部分。如果建筑业不仅可以在公司中,而且能在整个行业中应用重组,那么这将是促进重组的一个重要贡献,并将是建筑业的一个显著成就。不过,这个想法能否实现就得另当别论了,下文另有论述。

重组的目标

多年来,建筑业一直强调按时交工,在预算内和达到指定质量标准。这些目标与重组的目标非常吻合,那就是"在成本,质量,服务和速度这些关键业绩衡量上取得显著的进步",这正是建筑业最近才开始解决的。总体来说,重组的目标对建筑业来说并没什么新奇。翰默和珊比(Hammer and Champy)将重组提炼为四个关键词或特点:基础的、激进的、显著的、过程。这四个关键词为建筑业提出了一个有趣的问题。

重组的第一个特性是要求搞清最基本的问题"什么是一个公司的核心业务?"实质上这个问题要问的是"我们为什么在这里?""我们为什么以现在的方法工作?"对于"我们为什么在这里?"这个问题,企业的回答大概是"为客户服务"或是"为满足客户的要求"。遗憾的是,建筑业在这方面并没有良好的表现纪录。最近的一次在澳大利亚对建设公司的调查表明,虽然许多公司自称非常愿意得到客户的参与和反馈,但只有5%的受普查的公司做了正规的客户满意度调查。重组要求:没有任何事情是理所应当的,尤其是顾客。现状不是神圣不可侵犯的。当重组作为一种商业过程应用于建筑业时,它最根本的目标是施压于业界的竞争者,让他们真正地理解他们的工作的本质。

重组的第二个特点是它强调采用一个激进的方法,"激进"常常意味着"新的"或"独创的"。值得注意的是,英文单词"激进"源于拉丁文单词 radix,意思是根。韦氏(Webster)词典把"激进"一词定义为"与起源有关的",与"普通的"或"传统的"有很大区别。(翰默(Hammer)特别强调这点)。这种激进的方法用在建筑业时,可以提出这样的问题:"我们怎样进行重新设计商业过程,才能杜绝浪费?"而不是"我们怎样能够减少工地上的材料浪费?"在宏观层次上,思考重点将放在运用"节约施工"的原理和具体方法上,节约施工包括在整个设计和施工过程中减少浪费。

值得强调的是,虽然重组要求用一种完全不同的方法来定义问题和解决问题,但这并不意味着现存的管理措施完全不可取。虽然重组的精髓在于创新,但其他管理措施,包括全面质量管理、目标管理、平行施工、精细施工都是重组的基本组成部分。

对于翰默和珊比提出的重组的第三个特点,"显著的"是应用重组的特征,似乎在业界已达成了共识。成功的重组策略所带来的效果应该是突变性的改进,也就是大跃进。量的飞跃的概念好像只是对那些表现欠佳的公司会因重组从中受益匪浅。翰默和珊比反对这种看法,并归纳出三类公司将获益于重组。第一类是那些处于困境,为了存活,除了重组别无出路的公司。第二类公司目前状况尚可,但已出现可预见的困难。第三类公司是业界的佼佼者，无论在目前还是可预见的未来都不会出现危机，但他们仍然雄心勃勃，锐意进取。因此，业内没有任何一个企业对重组的显著影响无动于衷的。

然而，对于重组的显著作用，一个闪现于脑的问题是，"一个公司如何能够在一重复不停的突变性进步的状况下存在？"换言之，商业业绩一旦通过重组取得量的飞跃，回报降低的规律会不会也出现，其他对重组方法的应用会不会导致所得持续下降？莫利斯和布兰登已经强调了这个难题，他们将其描述为"变化规律性"。

"变化规律性是一个概念性的环境。公司一旦处于这种运行规律中，重组过程就永不会停息。随着公司改善，朝更优质和有效方向发展，重组也加速发展。这代表一个新的商业运作循环。"

这样，变化规律性回答了重组潜在的自我衰败的这道难题。因为认识到建筑业在未来将在一个新的环境内竞争，这个环境处于动态的变化之中，在这个环境中永恒不变的只有变化本身。

在翰默和珊比关于重组的定义中，第四个也是最后一个关键词是"过程"。其潜在的争论是大多数商业都不是过程型的。重组最关注的问题是再设计一套过程，而不是重新设计公司的部门或小组的结构。重组与这里提及的重新设计之间的区别是：它必须是根本上的，激进的和显著的。大多数使用商业过程重组公司，实际上也要从事过程再设计。对于大多数公司，这种方法代表激进的变革。但是这并不是翰默和珊比所指的"激进的"，搞清过程改进，过程再设计和过程重组之间的区别是很重要的（见图 3.1）。

过程改进包括最低限度的改变，相应的风险也低，同时对改进效果的期望也较低。过程再设计处于中间，有适度风险和高一些的期望值。过程重组是风险和回报都是最高的。

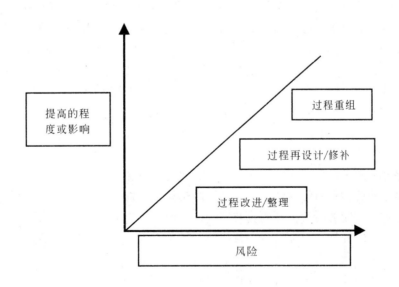

图 3.1　过程改进，过程再设计和过程重组的区别

这三个层次又可称为过程整理，过程修补和过程重组。过程整理是一种对现有的人、物和信息流动进行对号入座来减小浪费或重复。过程修补的方法是各部门用来发现工作中

的捷径或找出友好的工作方式。一般来说，过程修补不寻求改变整套程序也不寻求排除束缚。而重组是要排除一切物质和精神的障碍，为实现公司的目标重新建立一套方法。

重组方法论

初始阶段

重组是关于文化改变，而且经常是关于战略性和政治性的决定，它研究的是态度、行为和参考点的改变。它是关于改变态度、行为和参照标准的。因此，重组必须是由公司最高层开始的，是一个由上至下的方法。翰默和珊比(Hammer and Champy)观察到在实施重组的公司里常常会出现以下几个角色：

领导人：在建设公司里，通常是公司总裁或其他拥有同等权力的人，重组由他开始。

过程负责人：重组起始于公司总部，而不是在工地或项目层次。因此过程负责人一般是公司总部的董事或高级经理。

重组工作组：工作组会由接受重组概念的多个专业的成员组成，这些成员来自各个专业，从建设管理到信息技术，并且具有相当的技能和能力。

指导委员会：人数不多，由在公司内有威望的高级经理组成。

重组总管：翰默和珊比(Hammer and Champy)称之为"一个负责在公司内部发展重组技术和手段的人，并负责协调公司各个单独重组项目。"

杜宾(DuBrin)特别强调了过程负责人或项目经理的重要性，他指出这个人必须有权跨越各部门之间的障碍。

许多重组的例子都是从工厂基层生产环境中得来的，基层生产环境有明确的部门分工，这些分工不一定与建筑行业相类似。值得记住的一点是，在翰默和珊比(Hammer and Champy)对重组角色的定义里，没有什么会诅咒建筑业的内容，但我们必须清楚在建筑业中，我们讨论的是一个预先定制的产品(建筑物)，这个产品(建筑物)从一个组织中产生，而这个组织通常包括一群完全不相干的客户、设计者、施工者和使用者。这样，过程负责人在建筑业重组中，如果我们引用杜宾(DuBrin)的比喻，必须能够跨越一道道的鸿沟。

计划阶段

重组的好处来源于概念的通盘性，这一点看来已在理论者和实践者之间达成了共识。重组的重点更多地放在横向整体性上而不是纵向整体性，打破障碍，消除专家作用，重新划分界限。这种被实践者一致推崇的具有桥梁作用的方法论来源于系统方法。

传统的系统方法计划一个项目时，首先要通过建立系统界限来确定系统广义范围。然后才进而涉及细节。这种方法以表3.1表示(暗思壮 Armstrong)：

表3.1 常规系统法

概念上的	操作上的
(始于此)	衡量成功的指标
终目标	实施计划
各种备选方案	(终于此)

第三章 重组(Reengineering)

经典的系统法中的每个步骤都有一个独立的时间段并由时间反应步骤间隔。考虑每一个步骤时不参照下一个步骤。因此步骤的次序会出现反复应用。

系统思维的应用可以从表3.2中表示，它大致基于得克萨斯仪器(Texas Instruments)图表，该表列出了商业过程重组的组成部分。

步骤一"确定目标"的意思是开始于概念上的最高层次，并通过解决干系人利益"顾客需要什么"和"商业目标是什么"，来确定"业务需要是什么"。第一步应该在进入下一步"制定成功指标"之前完全解决。制定成功指标是系统法中很重要的一环，这必须建立在对"业务是什么"彻底理解的基础上。一旦前两步完成了，便可以在备选策略中作出选择商业过程重组或连续提高的决定。因而进入步骤四，制定方案。整个过程是动态的，步骤随需要按次序循环应用。

表3.2 系统思想的应用

步 骤	过 程
确定目标	确定顾客需要和商业目标
成功指标	成功指标是在对商业过程的了解的基础上制定的
考虑备选策略	选择：商业过程重组或持续提高
制定和选择方案	如果选择商业过程重组，则开始一套根本性改变的方案。 如果选择持续提高，则开始一套细微调整的方案。

COBRA(Constraint and Opportunities in Business Restructuring-an Analysis——关于商业再结构与机会的分析报告)项目制定了一个六阶段商业重组方法论。它与得克萨斯仪器(Texas Instruments)模式具有相同的特征。它也包括系统方法的循环应用。

实施阶段

如我们将在本章后面谈到的，在重组的实施过程中会有一些容易出错的地方。但是，我们暂时集中讨论有助于实施成功的一些因素。但是应该铭记的一点是"巨大的变革伴以相当的风险。"重组的风险随需要的时间来完成实施阶段并取得成效的增加而增加。我们需要证实实施过程正在一步一步地产生效益，这样才能让我们有信心继续实施下去。

库森—托马斯(Coulson-Thomas)根据自己的经验和观察，归纳出在实施重组过程中十五个成功要素。许多要素与重组中人的因素有关。在他看来，重组在本质上是对根本性变革的有效管理。他认为成功因素，比如高级经理和改革小组之间的相互信任是极其重要的；换言之，重组在很大程度上关心"情感、态度、价值、行为、决心和个人品格,比如思想外向。"

作为在建筑业中实施重组的先决条件，人/沟通因素的重要性也被默罕默德和亚提(Mohamed and Yates)反映在以下几个关键成功因素上(对重组是怎样应用于建筑行业感兴趣的读者，我们建议通读这本书)。

设计师，咨询顾问和承建商的对现行的设计和施工工作流程结构进行主要改变的高度的决心。

- 必须在主要的项目参与者之间建立并保持一个沟通的良性循环，因为信息交流有助于消除返工并因此缩短时间。

- 在工程的早期阶段必须寻求外部客户和内部客户的积极参与，以便在计划阶段掌握和实施他们的要求。
- 必须在施工过程的各项工作中制定和实施质量保证的技术。
- 在计划、签约、设计和施工过程中都应该鼓励创新。
- 应该研究新方法来提高建筑生产率。

所有的评论员一致同意重组的成功或失败关键在于文化转变的成功，而对于带来这种成绩的详细机制的看法却不甚一致。比如比多鲁和斯迪伯(Petrozzo and Stepper)就不同意翰默和珊比(Hammer and Champy)的观点，他们认为"重组项目的领导者不一定要全职"。至于采用何种管理形式为好也存在不同的观点。比多鲁和斯迪伯(Petrozzo and Stepper)的观点是"重组的领导人必须积极进取地与其他人沟通重组的重要性，重组将要做什么，重组什么时候会实行。"在某些人看来，积极上进地交流意见与库森—托马斯(Coulson-Thomas)的观点相去甚远。库森—托马斯(Coulson-Thomas)的观点更重视精神方面，他认为"组织是人们生活的集体。它们是敏感的有机组织体，反映我们的梦想和忧虑。网络正在不断地吸引知识工作者，在网络上他们共享资源"。

由于哲学和文化上存在差异，比如欧洲与北美集团之间，我们不可能给出一个明确的能够满足所有情况的重组程序。如果我们就比多鲁和斯迪伯(Petrozzo and Stepper)相对库森—托马斯(Coulson-Thomas)这一点举例说明，那么一些雇员可能会对硬性的独裁方法持肯定态度，而这一方式对另一个集团可能是灾难。概括重组的结果要比对重组过程进展中的每一个阶段步骤进行描述更容易些。但即使在这一点上，翰默和珊比(Hammer and Champy)说明，要对"重组商业过程是什么样子的"这一问题给出唯一的答案是不可能的。不过他们同意，总体来说有几点可以代表实施重组的结果，包括：

- 几项工作结合成一个
- 工人自己做决策
- 过程的步骤是以自然的顺序运行
- 过程有多种版本
- 工作是以最合理的方式进行
- 检查和控制被减少
- 调解最小化
- 一个项目经理提供唯一的联系点
- 混合集中/分散的运作较为普遍

上述特点是从北美企业的工厂层面的组织机构及生产程序中总结出来的。

默罕默德和亚提(Mohamed and Tucker)证明，在建筑业里，重组的应用可带来潜在的时间和成本的节省,(这篇文章因为对商业过程重组的研究完全放在行业的基面中(即建筑业)，而不像其他的文章在公司的范围内分析商业过程重组。因而特别有趣。我们建议读者通读这篇文章。)

潜在的时间节省

- 由于系统地考虑客户的需要，而制定出明确和准确的设计概要，因而减少变更和设计修改。

- 通过在设计阶段融入其他专业的参与,因而在满足客户需要的同时,制定出节省时间并有效率的解决方案。
- 通过与承包商协商更好的施工方法,实现共同认可的,具竞争性的总体施工时间。此承包商的选择是基于认可的过去工作业绩和财政稳定的条件上。
- 通过在设计阶段结合承包商的参与,提高设计质量。这将有助于现场的顺利运行及减少施工拖延。
- 在设计阶段全面考虑由于建设审批,材料供应和工地条件等约束因素从而缩短整体建设时间。
- 通过将小型工作队伍改组为大型队伍,提高分包商的效率,从而降低由于合作不利造成的工程拖延。
- 通过团队建设,合伙人制和策略性联盟的概念,增强项目参与者之间的关系来提高项目的业绩表现。这些概念把参与者之间的利益冲突降到最低,因而减少了项目延期的可能性。
- 通过减少工程材料和信息流动过程中浪费的时间,来提高施工效率。

潜在的成本节省

多数以上列出的时间节省的方法会直接或间接带来成本节省。除了这些由缩短建设时间而带来的成本收益之外,其它主要的潜在成本节省的方法总结如下:

- 制定一个能准确反映业主需要的设计概要,从而减少因设计修改,遗漏及相关拖延所造成的额外成本。
- 通过应用价值管理概念选择设计方案,从而避免选择具有相同功能的成本高的方案(业主得到最佳投资价值)。
- 运用平行施工概念并考虑项目设施的后期使用,使业主选择一个运行、维护和替换成本较低的设计方案(生命周期成本)。
- 基于过去工作业绩和财政稳定性选择承包商,能够在很大程度上加强控制财政和运作风险。
- 在关键的项目参与者之间实施团组建设的概念和在承包商和分包商之间实施合伙人制的概念,能够加强工作关系并减少高成本的冲突及争执。
- 在设计和施工过程中采用适当的质量保证措施,以确保减少最终会由客户承担的返工的成本。
- 采用一个高效率的材料管理系统,例如准时制("Just-In-Time"),以节省与材料运输、储存、偷窃、损坏等相关的运营成本。
- 以更加及时和准确的方法传递项目数据来减少以陈旧过时的信息为基础作出决策会导致额外成本的可能性。

重组的陷阱

重组的核心是系统方法以及它的整体性观念。系统理论的绊脚石就是如果将系统的组成部分从系统中分离并逐个进行优化,那么所得到的结果就是整个系统的优化方案。"在每一个具体过程里进行自我的商业过程重组,实际上减少了从整体角度进行改变的前景。

使企业现存的组织形式更加有效的结果是降低了对整体改变的推动力和渴望。"

因为重组的主要目的是打破传统的分组方式，比如，负责重组的经理有权利跨越部门间的屏障，因此分系统与整体系统之间的关系必然是重组实施者不断全神贯注的问题。商业过程重组中一个次优方案的例子就是商业过程重组鼓励和信任人们在工作中使用灵活和宽容的方法。但这可能与商业过程重组的为一个特定的任务确定方法的目的相冲突。举一个更进一步的例子，市场检验可导致一个企业被分成一些不同时间的契约协议，结果是没有企业来作为整体进行重组了。系统法对防止次级优化方案的自然反应是简单的持续扩大系统界限，以达到包容一切的目的。但是扩大的系统界限不一定总是一个实际可行的，至少是一个概念上理想的状态。

这并不是说重组理论有一个天生的概念缺陷，仅仅是指出重组由其自己导致了挑战。这个挑战如果不是颠覆性的，也至少是相当大的。大部分评论者都认为这是一个高风险和高收益的工作。

在实际操作上，翰默（Hammer）和珊比（Champy）列出了以下几个在应用重组时需要避免的错误：
- 试图去修改一个过程而不是去改变它
- 不把注意力放在商业过程上
- 忽视除了过程重设计之外的一切事物
- 忽略人们的价值观和信仰
- 总是愿意满足于一些微小的结果
- 过早地放弃
- 允许现有的企业文化和管理层的看法妨碍重组的启动
- 试图让重组从最基层开始
- 安排一个不了解重组原理的人作为重组领导
- 苛扣用于重组的资源
- 将重组工作埋没在企业日程安排之中
- 将精力分散于过多的重组项目中
- 试图在首席执行官还有两年就要退休时还进行重组
- 无法区分重组与其他商业提高计划的区别
- 将精力仅仅集中于设计
- 试图在不让任何人感到不快的情况下进行重组
- 当人们抵制重组带来的变革时就停止下来
- 分散努力

虽然以上列举的这几条是一套有益行动指南，同时也洞察到将重组引入一个企业时可能出现的陷阱，并且让人们清楚地看到那些不谨慎的人可能会遇到的陷阱。

对一家跨国的电子零配件企业在一个名为"聪明行动"的方案中引入重组的案例分析得出了以下启示：

"重组不是一项简单的或者自然而然的活动。从事重组时是不能够半信半疑或者措施不力。尽管选择需要重组的一个过程并不困难，然而要将其付诸实施却不是一件容易的事情。首先，必须选择一个正确的步骤和过程，这一步骤能够带来增值并被清晰准确地定义。必须授予重组工作小组权力,并找到执行决策的领导者。另外,要充分理解目前的商业过程,并清晰地、不模棱两可地表达出来是一件令人乏味的工作。而且通过谈判和执行计划来形成对重组的视野是有困难的,还会令人感到沮丧。重组工作小组或许会决定使用激进的变革,但是要管理层去承诺这种高风险的方案可能是非常困难的,尤其是去承诺一种改变现状的方案。"

与这些被察觉到的困难相反,该书作者注意到了重组过程具有一种玄妙的作用,这种作用虽然难以定义但的确存在。他们总结说"即使商业过程重组所要求的激进的改变是不可行的,商业过程的丰富也可能使得现存的过程在任何可能的时候增加价值。"这个总结性的评论似乎让人觉得有点遗憾。

信息技术和重组

正如重组被看作是授予现代商业能力来影响和进行激进变革,信息技术也被看作是允许激进变革受到影响的机制。很多人相信重组与信息技术的应用存在着内在联系。尽管"商业过程重组不是绝对地要求使用信息技术,然而商业过程重组的一个显著特点就是信息技术几乎总是在朱伯福(Zuboff)的术语中被用作'提供信息'去重建工作的性质。这是商业过程重组应用的关键特征。"在重组与信息技术之间存在着一种"推拉"的协作关系,即信息技术的力量扩展了重组的视野,而重组的挑战对信息技术的发展进步起着催化剂的作用。

信息技术可以被扩展为信息和通讯技术,这是一种有益的信息技术领域的扩展,特别是针对它与重组的关系时。好几位评论员在谈到信息通讯技术的发展与商业过程的关系时提到了温克吐曼(Venkatraman)的模式。温克吐曼(Venkatraman)模型(图 3.2)包括以下五个阶段:

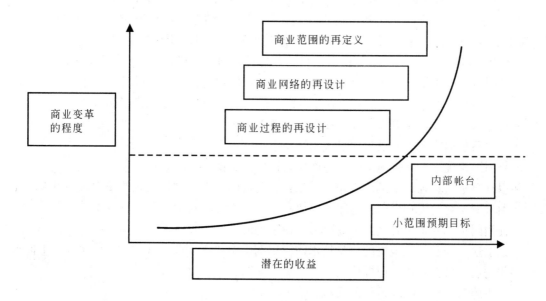

图 3.2 温克叶曼(Venkatraman)模型的五个层次

1. 模型的第一个层次被定义为"小范围内的预期目标",这一层次采用了标准的信息技术设备，对企业的组织结构做最低限度改动。这一层次代表了商业过程中个别的"自动化之岛"的不连续。
2. 第二个层次："内部整合"可以被认为是内部电子化建设的初始构筑，这种电子化建设整合了企业内部不同的任务、过程和功能，并利用这一平台整合了企业内部各部门的运作程序。温克吐曼(Venkatraman)认为，这是从"小范围预期目标"迈出的革新的一步。
3. 第三层次，也是变革的第一个层次——商业过程的再设计把信息和通讯技术看作是一个设计新的商业过程的杠杆，而不是简单地在现有企业结构上增加技术。
4. 第四个层次——商业网络再设计，包含了使用信息和通讯技术来使企业突破传统的自身界限，通过引入贸易伙伴间的更进一步的合作，将客户、供应商和竞争环境的变化包括进去。
5. 最后一个层次——商业范围的再定义是该模型的最后一点。哈斯特(Hazlehurst)提示了这个问题"信息和通讯技术在影响商业范围以及扩展了的商业网络中的商业关系方面发挥怎样的作用？"

温克吐曼(Venkatraman)模型对使用信息技术的商业的层次进行分类是有用的，特别是在区分进化和变革方面。哈森(Holtham)正确地强调了必须将作为商业过程的"过程"一词与应用于计算机科学和软件工程领域的"过程"一词区分开来的重要性。他指出，一些信息技术的拥护者们使用温克吐曼(Venkatraman)模型的第四个层次作为调整或仅仅是升级或取代现有的计算机系统的正当理由，而这当然是与温克吐曼(Venkatraman)模型的要义以及重组的哲学不一致的。

在建筑业中，应用重组的例子很少，但这并不是说没有信息技术用于重组活动的例子。尽管计算机技术的进步及其在建筑业中的广泛应用中开辟了新的领域，由于历史的、成本的和操作上的原因，这些应用仍然远远不能令人满意。默罕默德(Mohamed)和图克(Tucker)是赞同这一点的，他们将下列的信息技术确认为可用于建筑业重组：

- 计算机辅助设计(CAD)会议使得项目的参与者，例如客户、设计小组和承包商可以在不同地点、使用不同硬件以及计算机辅助设计(CAD)系统而协同工作。这一领域未来的发展将把设计工作室的计算机以及海外的建筑工地的计算机连接起来。
- 把电脑虚拟现实模拟用于计划和建造活动中。
- 把关系数据库用于一系列项目采购活动中。
- 知识库系统用于建筑施工规范与产品数据标准交换。
- 以计算机为基础对于工作流程的模拟势必使自动或半自动机器人的使用成为可能。
- 通过图形化的用户界面，运用了模块和有关知识库来自动计算工程成本。
- 使用图形化的用户界面的企业决策支持系统的发展。

以上这些处于领先地位的信息技术的应用例子主要用于项目计划和发展阶段。下面这些例子是有关模拟建筑工地操作的。

- 对于施工操作的统计过程的模拟。
- 运用图像模拟创造一个虚拟的工地环境
- 对于"精细工地"操作过程的模仿，来促进最大限度节省资源使用的技术的运用，比如准时制。

我们要注意一点，那就是由于科学技术所具有的特征，信息技术虽然富于魅力，但这种魅力可能不是增强了，而是诋毁了重组的作用。当许多企业在成功的重组的发展中使用了信息技术时，同样有许多企业在实现其重组的目标时很少或不注入新投资于信息技术。在这种情况下，这些企业的目标是：

- 不把信息技术作为关键途径
- 尽可能利用现有的系统
- 使用灵活的前端方法为一系列基本体系提供一个一般途径的范例
- 只有当过程已被重新设计好之后才进行发展，而且需要让所有参与的人都清楚了解新的信息技术的要求。

这似乎是避免信息技术在重组中滥用或误用的明智建议。在计算机应用方面存在一个在引进重组概念很久之前有的传统问题，究竟是要利用计算机使得现存的人工系统自动化，还是要利用新的计算机技术所带来的机会去重新考虑现有的工序。情况往往是这样的：首先将现有的工序自动化，然后在后一阶段意识到丧失了真正利用计算机能力的机会，接着又返回去完全重新考虑工序。在使用信息技术中，一条黄金戒律是只有对运作过程进行重新设计，新的信息技术命令才能得到重新发展。信息技术的一个重要作用是，它扮演着重组的促进者，而非唆使者。

欧洲人对重组的看法

读者现在也许可以总结出，重组与文化的价值关系相当密切。从其本质上来说，它是一个感性的概念。正如我们曾谈过的，我们不想加入我们自己的观点，而只想以一个中立的态度把当前重组理论的动向呈现给大家。在这本书中，我们总结了重组，因为我们相信，它是一种重要的现代管理活动，是值得建筑业认真考虑的。然而，假如我们不将一些欧洲的高层管理者对重组这一方法的影响和潜能所存有的严重的保留态度收录进来，这会是我们的疏忽。

西门子(Siemens)的领袖喜利起(Heirich von Pierer)发表于《商业过程重组》的文章中引述翰默(Hammer)激进的论文并且说，"我对翰默(Hammer)的激进的论文并不很满意。我们的员工不是中子，而是人。那就是为什么对话是重要的。"

哈森(Holtham)表达他的关于商业过程重组需要植根于特定的欧洲管理风格之中的观点时，谈到了对于欧洲思想中的人性的、全面的潮流的接受。这种欧洲思想与机械的、片面的美国方式是相反的，他宣扬促进企业内不同层次之间、企业之间、供给者和消费者之间以及民族与民族之间的合作这一概念。哈森(Holtham)对欧洲的商业过程重组的观点与翰默(Hammer)对商业过程重组在美国的应用的描述之间看上去似乎没有什么差别。也许表达信息的方式比被表达的信息本身的内容存在着更大的差异。正如哈森(Holtham)所总结的："商业过程重组的核心部分的价值超出了热衷于传道的北美方式。如果将这理论转换成适合欧洲情况的话，商业过程重组对于欧洲是有价值的。"

一个澳大利亚建筑业的过程重组研究的案例分析

没有一个统一的答案能够回答"重组商业过程是怎样的"这个问题。因为这个方法最根本的性质是依赖于创造性思维和对现存结构和障碍的破除，那么关于如何实施重组就不可能有一种特定的程序或列出的清单方式。然而以下这个案例分析的确体现了重组在初始、计划和实施阶段的许多特点。

（这一案例分析的材料是由作者从 T40 项目小组的最终报道中摘出的。感谢 T40 项目小组允许我们引用这些材料。有关 T40 的项目更为细节的讨论,读者请查阅爱尔兰(Ireland)的报告。）

背　景

这个 T40 研究项目是对澳大利亚建筑业的过程重组的研究,这一研究的目的是将施工过程的时间减少 40%。

表 3.3 列出了参与这个项目的企业。他们代表了三个主要的承建商,两个主要的材料和设备供应商,两个关键的顾问咨询公司以及澳大利亚国家研究和发展组织(CSIRO)。研究于 1993 年和 1994 年进行,于 1994 年 5 月公布结果。

表 3.3　参与的组织

澳大利亚 Flecther 建筑公司	澳大利亚国家研究和发展组织	BHP 钢铁公司
CSR 材料供应公司	A W Edwards	Stuart Bros. 公司
James Hardie 工业	Otis（奥迪斯电梯公司）	Smith Jesses Payne Hunt 公司
Taylor Thomson Whiting	Sly&Weigall	摩托罗拉(美国)促进员

研究方法

T40 小组由澳大利亚 Flecther 建筑公司的爱尔兰(Vernon Ireland)教授领导,每家参与的公司都负责研究过程的一个方面。研究基金超过了 300,000 澳元,其中最大的部分(96,000 澳元)来自澳大利亚建筑研究基金委员会。

鉴于美国摩托罗拉对他们自己的生产和项目管理活动的重组方面的经验,T40 小组雇用了摩托罗拉的人员来促进对于建筑过程的过程重组的分析。摩托罗拉也能够从一种与建筑业的传统不同的非建筑业的方法和角度来参与这个项目。

T40 小组进行了三个星期的集中的研讨会,他们研究了
- 作出"目前正在使用的"过程的流程图
- 重新设计"应该是怎么样"的过程
- 开拓解决方案的各个方面

小组的所有成员都对最后的报告有所贡献。这份报告反映了代表着设计和建筑过程的大部分参与者的小组成员的经验。目的是为了研究出一个再设计的过程以及一系列措施,而不是在现存的实践过程上加快速度。这其中的一些建议会容易得到实施,并且遇到很少阻碍,而另一些建议在实施前需要与行业中其他关键团体进行广泛讨论。

T40 项目的主题

本次研究结果包括许多关键主题:
- 将 T40 方案小组划分成为九个规模较小的小组,而不是成为一个由总承包商加上 50～100 个不同专业人士组成的大组,而且每个 T40 小组成员都直接阐述客户需求。

- 把一系列工作任务重新组织好以避免同一个专业人员对同一个建筑工地进行多次访问。
- 保持对业主负责的态度。
- 对负责实施的九个小组(包括建筑师和结构工程师)给予经济上的奖励和惩罚,来促进他们集中于具体行动中。
- 商业活动建立在信任与公平交易基础之上,从而消除了检查其他企业的检察员的工作的需要。
- 第一次就做好,避免返工。
- 取消传统的招标方式,从而消除了招标的时间和成本,同时使方案解决小组直接满足消费者的需求。
- 方案解决小组实行资源共享,而不是重复功能(如计划、监督、以及雇用有执业资格的分包商)。
- 管理层和劳工层之间合作而且组成团队。
- 与当地有关政府审批部门建立合伙人制的合作。

T40 小组意识到,改变关键参与者的态度以实现创新并不容易。业内许多人会说这些目标是不可实现的。然而该小组相信,这些改变对于澳大利亚建筑业的发展是必要的,并且也可以在一段时期内实现这些目标。这些目标的实现将把澳大利亚推到一个在亚洲庞大的建筑业发挥战略性作用的关键地位。

T40 过程的方案

在顾客与交货小组之间达成共同的目标

T40 项目的方案是各个问题解决小组的所有成员直接为顾客的需要而工作,而不是让这些需要由建筑师过滤一遍,再通过总承建商过滤一遍之后再来做。整个 T40 小组将直接集中于提供一个能够增加顾客商业价值的解决方案。

顾客和方案解决小组(T40)的成员们都需要说明他们想从过程中得到什么以及满足其他人的目标的各种限制。为了实现这一方法,需要对现状做出以下改变:

- 必须发展互相依靠和彼此信任的新方法,加之经济上的奖励去支持整个小组作为一个整体去寻找项目来加强这种方法(分享成功的奖金和惩罚)。
- 必须发展新的反映相互依赖的工作模式,例如接受传统工作内容的重新分配。
- 业主(顾客)与参与小组间必须在公司的最高层的合作达成约定和承诺。

简单化了的过程

该 T40 项目是一个设计连建造的单一阶段的过程,在这个过程中方案解决小组从最开始确定顾客的需求的时候开始。关键问题是使得顾客因物有所值,并在约定的缩短的时间之内交工而满意。

T40 研究小组基本信条是,由 50 到 100 个独立承建商组成的建筑队的长期效果是不如拥有 8 到 10 个关键参与者的建筑队的效率高。控制着 50 至 100 个分包商和供应商的项目经理只能救一时之急,而没有时间和精力顾及创新的机会。

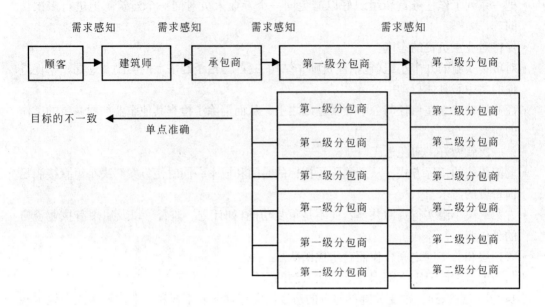

图 3.3 传统的过程的等级层

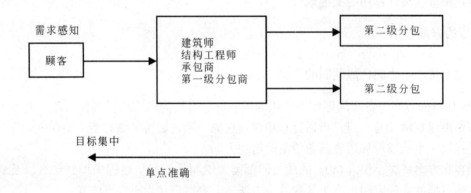

图 3.4 T40 的解决方案

T40 的建议是,在一个典型的高层办公楼的建造过程中,一个承建商应该负责以下的每一条:
- 工地现场办公室的建立(主承建商)
- 土方开挖(主承建商)
- 主体结构施工(主承建商)
- 幕墙分项(主承建商)
- 除电梯外的所有设备(设计和安装)
- 电梯(设计和建筑)
- 内墙间隔

- 混凝土的供应
- 钢材的供应

建筑师和结构设计师也是小组的成员，他们也是项目目标的主人，也需要分担适当的风险和分享适当的报酬。

在内部装修阶段，会有两个设备承包商和两个装修公司参与。尽管这种安排确实引入一个三个层次的等级的装修和设备安装合同，但是关于对这些专业小组没有控制范围和业绩奖励的传统理由是能够被时间和地点上的分隔所克服的。

在两个分包商之间的工作内容应该有着一个清晰界定，来消除互相干扰，并合理分担风险和报酬是必不可少的。

将在分包商中出现的创新包括：
- 创新的系统的发展。
- 在第三层次的分包商间更统一而整合的行动。
- 对子过程的重组，以消除协调中的问题。
- 对过程中的主要成本内容进行分析。
- 主要成本内容与顾客所期望的价值之间的关系。
- 每个建筑工人的活动的更详细计划、分工与统一。

这种安排的潜在缺点可能是在协调方面的问题和困难，一个中层组织结构会使主承建商与具体做这项工作的分承建商分开。理想的情形是，每个设备和装修承建商直接雇用工人。然而，假如他们不能够这么做，由于工作量的波动，对于三级管理结构中的中间层次的承建商必须有奖励，使他们创新并通过他们的协调来增加价值。

施工过程重组的实例

（这是 T40 研究报告给出的三个例子中的一个）

三位电工

在目前的安排下，由不同分包商所雇用的三个电工有可能在同一个顶棚的空间下的相邻的梯子上工作。一个在安装电灯，一个在安装烟雾测试器，还有一个在安装空调。他们可以各做 30min 的工作，然后开车去下一个项目，其中也包括类型电工工作等等。一种更为有效的方法应该是每一个电工完成一个项目中的所有电工活。

表 3.4 和表 3.5 表示了这一逻辑的关系，这两个表列出了分包商的传统安排以及 T40 所建议的分包商的一整套工作安排：

表 3.4　分包商的传统的工作安排

活　动	现在的活动(任务)	由谁做	访问次数
1	空调管道	A	1
2	给排水埋管	B	2
3	洒水器水管	B	2
4	铁钉固定隔板结构	C	3
5	电线吊篮	D	4

续表

活动	现在的活动(任务)	由谁做	访问次数
6	烟雾测试器埋管	B	5
7	电路埋管	D	6
8	计算机线埋管	E	7
9	电话线安装	F	8
10	顶棚吊顶龙骨	G	9
11	洒水器头、空调控制器和电灯安装	G	9
12	空调安装	G	9
13	洒水器安装	G	9
14	石膏隔板	G	9
15	铝合金支架	H	10
16	玻璃隔板	H	10
17	细木工	I	11
18	框和门扇安装	I	11
19	隔板的第一层油漆	J	12
20	墙裙管道	K	13
21	地毯和塑胶板	L	14
22	计算机线路安装	E	15
23	电话线路安装	F	16
24	电线安装	D	17
25	烟雾测试器安装	B	18
26	顶棚吊顶安装	G	19
27	隔板的第二层油漆	J	20
28	计算机工作站的安装	K	21
29	测试电源和灯具	D	22
30	测试空调设备	A	23
31	测试水管和洒水器	B	24
32	测试烟雾测试器	B	25

表 3.5 T40 所建议的分包商的工作的安排

活动	T40 所建议的活动	由谁做	访问次数
1	空调管道 供水和洒水器埋管烟雾测试器 电线吊篮 电话线埋管 电线埋管	A	1
2	计算机线路埋管	B	2

续表

活 动	T40所建议的活动	由 谁 做	访问次数
3	顶棚吊顶龙骨 洒水头、空调控制器和电灯的吊顶板 隔板 石膏板 铝合金框 玻璃 细木工 门扇 第一层油漆 墙裙管道	C	3
4	计算机线路	B	4
5	机械的和电子的安装	A	5
6	电话安装	D	6
7	顶棚格板、隔板的最后一层油漆	C	7
8	安装工作站	E	8
9	测试并启用设备	A	9

劳工授权

在制造业的许多部门以及建筑业的一些部门中存在着很强烈的动议去重新定义领班这一角色，并将其传统的计划者和监督者的角色转到教练的角色，从而鼓励工人对他们自己的工作内容、工作计划、工作中需要的材料的采购和运输，以及相关活动负起责任来。劳工授权的具体方面是授予以下这些工人的：

- 充当领班在确定和纠正违反安全措施的做法和质量保证问题的眼睛和耳朵；
- 订购他们自己所需的设备和材料；
- 由负责解决工地上任何潜在纠纷的工会所代表；
- 负责把他们自己发展成已协定的工作结构。

劳工的团队合作

在T40项目的环境中所寻求的是实现一种基本的文化上的变革，这种变革使得管理层和普通劳工理解并尊重每个人或者每个团体的需求和随之而来的目标。

T40项目的解决方案包括了以下团队合作概念：

- 鼓励参与的公司制定企业间合作的协议，这种协议可以是作为一个被认证了的协议，或者是一个关于需求、目标和报酬的正式的承诺和谅解备忘录。
- 树立不仅代表各干系人的，而且代表T40项目的参与者的目标。
- 将这些目标传达给T40项目中的所有参与者。他们将经常确定那些普遍的而且被确定为与他们的企业相关的东西。

- 制定用于评价共同目标的方法。
- 鼓励工人顾问委员会和智囊团会议来就已实现和未实现的目标进行经常性的交流。

共同决定

合作的延伸就是共同决定。共同决定的一个基本的原则是解决项目周期内的工业关系问题时不涉及损失时间的绝对的承诺。

人们的创新

人们希望那一群被授权的人对于项目的创新能够做出重要贡献。给人们授权的一部分目的是让他们释放出和谐和力量，因而鼓励他们将项目看作是他们自己的，这样他们就会为项目的思想库做出贡献。这样的安排可能会成为一个项目合作协议的基础或是公司合作协议的附录。

与地方政府成为合伙人制的合作

T40项目建议的基本主题是与当地政府成为合伙人以同时满足政府和申请人的需要。这一建议的某些内容可能会被立即实施，但其他内容将需要较长期的变革才能实现。

较长时期的计划改变

一个较长时期的计划审批解决方案是针对建筑物的使用和外型而言的而且与土地规划有关，假如项目的方案符合有关的指导方针，那么政府有关的审批部门就会同意该计划。

投　标

T40项目过程的中心假设是，整个决议小组，包括总承包商和专业分包商都完全致力于满足顾客的需求。这一过程的重要的要求与包干的投标是很不相同的，它甚至与当前的由外部顾问咨询公司进行基本设计然后在这一基本的设计和要求的基础上进行招投标设计连施工的方式也很不同。目前正在使用的这两个方案将决议小组排除在直接表达顾客需求之外。

T40所建议的这一过程对于承建商来说避免了现有的，成功率约为1/10的投标的浪费。在这一系统中，需要以下面几条为基础来选择承建商：

- 通过查阅行业在一个时期的数据库来了解该承建商在时间上的过去表现的记录。
- 建立在完整的设计基础上的共同同意的造价。
- 第三方认可的与最佳价格相近的造价。
- 双方都同意的远远比同类建筑物的平均施工时间短的施工时间（如时间缩短40%，相应地在成本上缩减25%）。
- 对公共社会负责。

这些要点列于表3.6。

表 3.6 传统，文件和建造与 T40 项目的解决方案之间的区别

	完整的文件包干合同	文件和建造	T40 过程
总承包商参与对顾客需求的确定	无	无	有
分包商参与对顾客需求的确定	无	无	有
完全责任	有	有	有
分包商和供应商的数量	50~100	50~100	在第一层次以 8~10 个为一组
对分包商的奖励和惩罚	无	无	有
认为分包商与业主有关联	无	无	有
明确项目批准的不同阶段	有时	有时	是
快速设计	无	有时	有
设计和施工过程中的变更	通常很多	中等到少量	无
文件整理过程	很多不同的原始文件	有一些协调	阶段性的，分项的和完全的
项目沟通	电话和传真	电话和传真	直接
专业资源的利用	分别	分别	共享
监督	分别	分别	共享
自动开出付款清单/要求	很少	很少	有
投标	占 6%~8% 的成本和超过三个月的时间	占 3% 的成本和超过三个月的时间	几乎零成本
当地政府部门批准	6~12 个月	6~12 个月	0 个月
可预见的结果	无	较好	有
要完成的总的工作量	W	90%W	75%W
时间结果	T	90%T	60%T

结 论

在这一章中，我们追溯了重组的起源，验证了它的目标，并详细说明了这一在商业领域普遍为成功的重组实践者所接受的策略。我们还详细叙述了 T40 项目，这一项目作为在建筑业中的过程重组案例分析，代表了重组方法的主要特征。

我们将北美重组方法的直接性与较为克制的欧洲观点进行了比较。我们自己关于重组的观点是一种谨慎而积极的态度。当前在建筑业中成功运用重组的例子还很少，因此我们无法对它的未来更加乐观。虽然这么说，但我们确实注意到了，澳大利亚联邦科学和工业研究组织(CSIRO)、房屋、建造和工程部门开始了一个旨在为建筑过程的重组提供解决方案的为期三年的研究计划，这说明在建筑业界有一股接受重组原则的浪潮。引用下面这段话：

"澳大利亚联邦科学和工业研究组织(CSIRO)致力于建造过程的重组,并意识到重组能够有效支持澳大利亚建筑业在国内和国际市场上对抗海外的竞争。"

重组能够改变建筑业的结构和文化。时间会告诉我们该行业是否会迎接重组方法所带来的挑战。

第四章

合伙人制(Partnering)

导　言

　　"合伙人制"是很难定义的。它对不同的人有着不同的意义。合伙人制与人际关系，干系人利益，以及权力平衡有关。换言之，合伙人制与人与人之间的互相作用有关，而因此产生的不可避免的结果就是，它是一个难于说明和分析的复杂题目。合伙人制不仅仅是旧时尚价值的正式化，或者怀念过去"君子无戏言"的好时光(虽然道德责任和公平交易是任何合伙人关系的基础)。它也不仅仅是一项建筑采购技术(虽然建筑采购技术可以被用于良好行为的操作化,带来文化转变,因而建立一个更加团结的团组)。合伙人制在建筑行业内的应用是被大力提倡的，而且很多人认为可以成功。学术杂志期刊中关于合伙人制的文章题目洋溢着乐观积极的信心。比如"合伙人制意味着结交朋友而非敌人"，"合伙人制 使人得益"，"合伙人制有意义"，以及更有说服力的"合伙人制——90 年代的唯一方法"，诸如此类的题目大量出现在专业期刊上。

　　本章中我们将本着实事求是的原则去探导合伙人制的起源，讲述其是如何被建筑行业采用的，以及合伙人制在实施过程中的优势、风险和缺陷。

合伙人制的起源

　　合伙人制作为建设管理的一个正式概念起源于相对近期，可以回溯到 20 世纪 80 年代中期。这并不是说在此之前并不存在合伙人制的做法，而实际上很多人都会说"私人业主和承建商的合伙人制与建设本身一样古老。"曾有人指出，在英国，象波维思(Bovis)这样的大型公司在 20 世纪 30 年代就与相关业主发展了友好关系的文化和传统。

　　本章的讨论将集中在 20 世纪 80 年代后也就是当"合伙人制"有完整的内涵之后。因此，我们将集中讨论正式的合伙人制，也就是说合伙人之间有完整的合作安排的证据的情况。当然这并不是要与非正式合伙人制的存在和重要性进行辩论，只不过我在此不考虑非正式的合伙人制。

　　根据国家经济发展办公室(National Economic Development Office)的报告《合伙人制：在没有摩擦的情况下签订合同》，正式的和真正的合伙人制只是在 20 世纪 80 年代中才正式建立。而且第一次的合伙人制是壳牌(Shell)/帕孙(Parsons)/斯泊(Sip)在 1984 年的合作。而在 80 年代期间最经常被其他文章引用的是杜邦(Du Pont)和福乐欧·丹尼尔(Flour Daniel)在凯普维牙(Cape Fear Plant)项目的合作关系。这个合伙人制合约在 1986 年签定，其实是对 1975 年以来这两家公司一直就已存在的关系的正式化。这个时期的其他一些合伙人制关系包括卡比德(Carbide)和贝喜特尔(Bichtel)，普罗特(Protor)和甘波尔(Gamble)/开乐(kellog)和壳牌(Shell)石油/帕孙(Parson)。

建筑业内的合伙人制

大部分评论员认为,合伙人制在建筑业的应用和出现归因于20世纪80年代美国建筑业协会(Construction Industry Institute of the United State——CII)的工作以及美国军事工程部队(US Army Corps of Engineers),采取合伙人制(主要通过查尔斯·可文(Charles Cowan)的努力)在现阶段,在美国已可以找出很多合伙人制的实例,而且在新西兰、澳大利亚和英国等国也开始在这方面开展工作。在澳大利亚,合伙人制的采用部分源于盖力斯(Gyles)的皇家委员会的新南威尔士州建筑业生产力的调查报告;在英国,则部分由于拉谭(Latham)的报告的倡导。拉谭(Latham)在其报告《相信团队:建筑业合伙人制的最佳实践指南》前言部分提到"合伙人制可以改变人们的态度,改进英国建筑业的业绩。我希望此行业及其委托人现在可借助这份报告着手实施合伙人制。"

在澳大利亚,盖力斯(Gyles)的皇家委员会的工作则更进一步,他们对合伙人制进行了初步研究,将其作为鼓励新南威尔士州建筑业文化转变的一种手段。合伙人制是澳大利亚新南威尔士州公共工作服务厅承包人委托计划中的认证标准之一,这说明了一些政府委托方对于实施合伙人制的重视程度。

对建筑行业采取合伙人制的程度目前还很难进行定量分析。然而在美国有众多的成功的合伙人制的实例。在1994年对2400名律师、设计人员和承包商的一项调查中,项目合伙人制和调解被列于解决争端办法的排名表的首位。美国建筑业协会(CII)在1994年波士顿举行的年度会议中报告:"在工程安全,可建设性,全面质量管理及短期和长期合伙人制中,长期合伙人制所带来的节省是令人难忘的。"美国建筑业协会(CII)策略联盟任务组报告指出,使用长期合伙人制的196个工程中,平均节省了15%的安装成本。对于以单一工程项目为基础的合伙人制来讲,美国建筑业协会(CII)的团组建立任务组报告指出,有五个大工程项目的造价平均节省7%。这些统计数据暗示着合伙人制在美国的普遍接受程度。

在澳大利亚,主要建筑商协会(Master Building Association)每年都会举办一次竞赛活动,这些竞赛吸引了很多以合伙人制进行的样板工程项目参加。但赫拉(Hellard)对于合伙人制在英国的实施程度有些质疑。"目前还没有一个完全的包括委托方、设计单位、总承包商和分包商并且有一份正式的工程计划书和涵盖正规培训和义务承担的合伙人制项目。"不过,一些关于英国合伙人制的案例分析已被收入在最近出版的《相信团队:建筑业合伙人制的最佳实践指南》的报告中。

合伙人制的目的

在本章开篇处我们已经说过,合伙人制是一种很难独立定义的现象。与之相比,定义其目的则要比定义合伙人制的属性简单的多。

目前流行的有关合伙人制的目的的定义就有许多。其中一些定义十分广义,例如"合伙人制是改善参与工程项目各方之间的关系,使参与各方都能受益的过程"。另外一些定义则较为具体,但仍包含相同哲学,比如:

"合伙人制不是一种合同,而是对于合同中体现出的良好信誉的认可。合伙人制试图通过一套在义务承担及信息交流方面发展为较完善正规的方法策略,在干系人间建立起和谐的工作关系,并创造出一个环境,用信任和团队的合作防止怀疑和争吵,培养一种利己利人的合作关系,并促进一个工程项目成功实施的工作环境。"

所有的评论员一致强调达到信任和合作是合伙人制的基本目的。典型的描述就是"合伙人制的目的是两个或多个以信任和合作为基本精神的团体间的长期的契约性承诺。目的是让每一个参与者都能最大限度的发挥他的潜能,并不断提高其业绩水平"。

可文(Cowan),现代合伙人制运动中的主要代表之一,曾强调说,"合伙人制不仅仅是一套目标和实施步骤;它是一种精神上、心理上的状态,是一套哲学理论。合伙人制体现出了对于参与的团体双方的所有干系人间的尊重、信任、合作以及优势互展的一种承诺。"

合伙人制类型

合伙人制可以分为两大类型,这其中又有许多种类。合伙人制的两个类型是:策略合伙人制和项目合伙人制。(策略合伙人制有时也指"多项目合伙人制"或"第二层次合伙人制",项目合伙人制有时指"单个项目合伙人制"或"第一层次合伙人制"。)策略合伙人制应用于当两个或多个公司在长期基础上采取合伙人制来承建多个建筑工程项目的情况。项目合伙人制与此相反,发生于当两个或多个公司通过合伙人制,共同承建单个工程项目时。在美国,90%的合伙人制是项目合伙人制。因为合伙人制是关于一种长期合作的关系,因此,与短期的项目合伙人制相比,长期的策略合伙人制可以带来更多收益。但是项目合伙人制可以成为策略合伙人制的第一步。

[注释:鉴于本段文字内容,我们使用"项目"一词来表述那些需要正常的施工时间的工程项目。之前我们曾将杜邦(Du Pont)/福罗欧·丹尼尔(Fluor Daniel),卡普·非亚(Cape Fear Plant)工厂等项目称作早期的合伙人制协议。虽然这只是单个项目,但它却有空间上的和时间上的多重维度,因此将其划入策略合伙人制范畴,而非项目合伙人制,则更为适宜。在大多数情况下,"项目"和"策略"间的差异是一目了然的。]

项目合伙人制

参与者

对任何一种合伙人制,首先要解决的问题就是,谁应该参与合伙人制的安排。这立刻就引升到"鸡和蛋"的关系上。如果合伙人制概念是在项目一开始时被引用的,那这就会影响所使用的项目采购方法,而此采购方法又反过来影响项目参与者的组成。为了获得合伙人制的最大益处,普遍趋势是使用设计与建造(在澳大利亚称为设计与施工)这种采购方式。这种方法的好处是,它让所有的主要干系人,例如业主、设计连建造承包商(他们的公司也有设计小组),都能在项目一开始的时候就参与合伙人制的安排。

然而合伙人制并不只应用于设计连建造的项目采购方式,它同样可以应用于传统的采购方式的工程项目,即在接受投标阶段评定最低标价中标的承包商。在这种情况下,承包商被排斥在设计阶段以外,因此在项目筹备阶段,合伙人制只涉及业主和设计顾问公司,而承包商是在投标接受阶段才参与合伙人制关系。这并不是实施合伙人制的理想环境,就好像在比赛过程中将一个新球员插入到足球队联盟的基层阵容中,并期望他们可以成为联赛冠军一样。

合伙人制团队成员的选择要建立在完全信任和相互支持的基础上,这点极为重要。合伙人制团队必须有能力在整个项目过程中承担起其全部责任。"如果团队之中任何一人没有担负起他的职责,那么任何合伙人制过程中的尝试都将宣告失败。"因此选择合伙人制的团队成员并不是一件简单的事。正像可文(Cowan)说的,"理想情况下,你希望选择那些在以往合同中有过成功合伙经验的承包商或业主,通过对承包商的面试,业主可以了解其兴趣和专长。"初看这段话会觉得有些语义重复。但是对于承包商而言,确实存在着从兴趣方面进展成为合伙人制参与者的可能性。这一转化过程在新南威尔士州承包人资格鉴定计划书中有所体现,其中从低到高定义了五种合伙人制实施方式。参见表4.1。

表4.1 合伙人制特征的等级

等级	阶段	特征
低级	认可阶段	认识到合伙人制的益处
普通	发展阶段	发展合伙人制的指导策略
较好	建立阶段	承诺在所有工程项目中采用合伙人制
好	持续提高阶段	通过合伙人制来不断提高项目实施过程的已存档的证据
高级	最佳实践阶段	设计顾问、供应商和分包商之间的长期合作关系的记录

承诺

由于一方必须要信赖另一方,因此选择合伙人的决定是非常重要的。因此这个决定应该由企业组织的最高领导阶层来决定。"不可想象……当两个企业组织的最高层没有一致的承诺,双方怎么可以实现合伙人制。"

虽然所有的评论员都着重指出高层承诺的必要性,但在实现高层管理者的承诺,达成合伙人制协定之前,仍有必要在企业组织内部进行"内部合伙人制"。"内部合伙人制"是指使团体准备就绪、整理回顾内部的各程序步骤和文件。它还包括教育和知会公司员工有关合伙人制的事项等工作。其结果就是建立一个统一的企业,以达到与其他企业密切合作的目的。

合伙人制过程

合伙人制过程必然包括多种形式和变化。在本节中我们将讨论单个项目合伙人制,其过程将与策略合伙人制有所不同,后者我们会在稍后段落中提及。项目合伙人制的实施过程将依据各项目的不同情况而定。如果项目参与者在此之前已经加入了类似的合伙人制协议,那么问题就好办多了;反之,如果参与者是第一次加入此过程,那么该过程的发展则需要特别的留意。本节中我们将假设我们面对的是初次合作者。

基本上,合伙人制过程是团队建立的过程,这就是为什么内部合伙人制获得成功的结果的作用显得至关重要。合伙人制过程必须被看作是达到目的的一种手段而非目的本身。凯森白(Katzenbach)和史密斯(Smith)指出,"实现优秀业绩过程中的各种挑战将造就一个团队。对于团队的成功而言,对实现杰出业绩的渴望远比团队建立本身,特别是奖励条例或者领队的理想条件重要得多。"实质上,团队是被来自合作中的各种挑战所激励,而这种激励的产生和保持又是合伙人制过程的主要功能之一。

一般而言,合伙人制过程有三个阶段,即:项目前期阶段、执行阶段、完成及回馈阶段。

项目前期阶段

项目前期阶段开始于决定是否选择合伙人。有些情况下，一些政府招标的预选资格条件就是使用合伙人制。无论合伙的最初愿望是否来自业主，设计小组或是承包商，它都是一个潜在的高风险决策，因而不能轻而易举地采用。合伙人制也并不总是必然的最好的方法。对有些当事人和有些情况来说，最好避免使用合伙人制。理想状况是，每一方都能有互补的长处，而且在这些当事人之间有一种和谐关系。最糟糕的情形则是，某一方内在的弱点散布到整个合伙人团组中，在这种情况下，合伙人制非但不能稳定局势，还很有可能恶化局势。

假设已决定采用合伙人制，那么如何开始选择合伙人呢？对于这个问题没有明确的答案。在某些情况下，例如澳大利亚新南威尔士州公共工作服务厅就将合伙人制作为政府工程项目招标的资格认证的前提条件。然而，虽然有新南威尔士州公共工作服务厅的鼓励，合伙人制中的(各层)联系仍须由独立团体构成。在另一种情况下，合伙人制的初始决定来自于团体内部的个人。一些奖励计划，例如澳大利亚建造商合伙人制奖励，以及诸如《相信团队：建筑业合伙人制的最佳实践指南》等报告对合伙人制的实践起到了播种的作用。即使合伙人制不得不做到信任及相互之间的合作，但在最初阶段，可能合伙人之间的提案进程仍是不透明的，甚至在某种程度上作为一种商业机密保持着神秘色彩。下面是一个由业主向一家设计连施工的承包商提议合伙人制的例子，可以让我们大致了解到合伙人制的实施是如何开始的。

"为了最有效地完成这项合同，业主计划与总承包商和分承包商建立一套有凝聚力的合作关系。这种关系将有助于利用每个团队的优势和力量，以期在不用返工的情况下，一次性在预算内按计划建设出高质量楼宇。这种合作将是双向构成，而且是自愿参与。任何与实行此合作有关的费用都必须经过合作双方同意和平均分摊，而且原合同价不变。"

由此可见，这是自愿性协议，目的在于创造一个相互信任和抵制对抗性氛围的合作文化。(到此阶段，读者可能会感到，传统的建筑工程合同的根本属性就表现了合伙人制目的与正规合同协议书之间的相对抗性。这是一个较难的问题，我们将留待后面有关合同、法律问题的一节中探讨，现在暂时将其搁置一旁。)

在高级管理层之间对于进行合伙人制达到了共识之后，合伙人制的发展进程就走进了一条更可预测的道路上。图4.1是合伙人制实施过程的模式流程图。

合伙人制第一次研讨会

一旦合伙人制决策产生，就要把负责日常工作的主要中层经理组织在一起。在设计连建造项目采购方式中，这个小组由来自业主，设计小组和施工小组等干系人组成。除此之外，或许还有法律及金融顾问，有些时候还有来自于法规计划和建筑管理等政府部门的代表。为了达到小组的活力，小组规模不应太大，应该限制在最多25个人左右。工作组通常在首次研讨会被相互介绍(此研讨会的详细机制将根据具体的情形和不同国家的文化背景而有所不同。)首次研讨会的目的是为今后的发展制定基础章程，并且应在合同签定后尽快举行。通常情况下，研讨会的地点应该选在远离所有干扰的隐蔽而居中的场所进行，会期可以是一天、两天或者三天，根据项目具体情况和参与者对合伙人制的熟悉程度而定。研讨会期间最好有专人管理，其中至少应有一部分的会议时间由一位有合伙人制经验的促进员负责管理和协调。

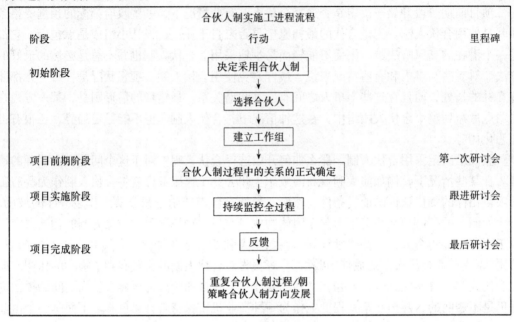

图 4.1　合伙人制过程的模式流程图

　　这首次研讨会的目的是：公开交流，建立团组精神，建立合伙人制目标，建立一个实现此目标的计划，获得与计划一致的承诺。研讨会的通常形式是以团队建立作为开端，并鼓励参与者发掘彼此间性格上的共同点。这也许会低调进行，或者采用更为正式的方法，例如使用迈尔斯—波丽格斯(Myers-Briggs)型的指示器将人格特点进行扼要说明。接着，工作组将发展出一套概念性框架或者描绘出大家心目中对于合伙人制的理解及它将如何影响各位参与者在此项目中的工作。工作组进而会按照 SWOT 分析法考虑并认同彼此间的目的和利益所在。(SWOT 分析法即指对于 S 优势、W 弱点、O 良机和 T 威胁进行分析的方法。)共同的目标如时间、造价和质量会是主要的议题，同时其他诸如安全、可建设性、电子数据交换等问题也会在研讨会中讨论。实现这些目标的方法通常是在参与者中建立一个行动小组，处理那些在研讨会过程中出现的问题。这些行动小组由各合伙人中的相应专业人员组成，例如法律顾问，或者包括一系列不同专业人士，如建筑师和工程师。不管是其中任何一种情况，行动小组都将担负起为整个团体提供解决问题的办法的职责，这些问题包括找到业绩指数、纠纷解决和减少的办法，以及安全措施等。行动小组协助整个团体的结合进程，并将合伙人制中的问题减至便于管理的程度。多数情况下，行动小组的成员将在整个项目实施过程中保持密切的交流与联系。

　　首次研讨会最终要达成合伙人制章程协议和合伙人制执行计划。这次研讨会也应该就跟进研讨会的时间安排达成协议，这是为保证合伙人制成功实施所必需的。行动小组也需要为跟进研讨会担负起一定的职责，即发展各自负责的题目，跟进研讨会是每隔三个月举行一次，贯穿整个项目周期。

第四章　合伙人制(Partnering)

合伙人制章程(或者项目章程)通常是由合伙人签署的。(由所有参与人签署的章程被看作是一种很具象征意义的行动,它是对彼此表示忠诚,公认个人利益低于项目利益的一种实际行动。)章程的签署者(以设计连建造的项目采购方式为例)将来自于如下团体:业主、承包商、设计工作组、分承包商、厂商、材料供应商、咨询顾问方(如法律顾问和规划专家)。

所签章程也许是非常概括性的,也许很详尽,这取决于合伙人的决定和项目的特性。假设该章程的复印件将被贴在所有合伙人的墙上,并且在从业主的办公室到工地车间和工地办公室等不同地点展示,那么一张 A4 大小的纸就是限制条件。图 4.2 是一个合伙人制章程实例:

合伙人制章程

我们,××××项目的合伙人承诺在相互信任、共同承担的环境中一起工作,并为此项目中所有干系人最大利益而努力。

此工程项目的目标是:
项目按时竣工
成本控制在预算之内完成项目
取得合理的利润
实现由事故引起的时间零损失
建造零缺陷的高质量楼宇

我们团队的使命书就是:
我们将在彼此开放的沟通的环境中合作,
为实现××××项目的优秀做出承诺

签名:

执行阶段

在第一次合伙人制研讨会结束的时候,会建立一套包括一系列跟进研讨会时间表的执行方案。这些跟进研讨会的次数和每次开会的时间长短取决于每一个项目的具体情况。第一次研讨会是为了实现从传统的对立性的文化,向双赢环境的团队精神的转变。这个人际关系的量的飞跃需要不断地培养(尤其是对合伙人制的新来者)。跟进研讨会对于加强合伙人制文化是非常重要的。跟进研讨会通常延续与第一次研讨会相似的模式。如果在第一次研讨会时使用了外部促进员,那么在跟进研讨会中使用同一名促进员也是合情合理,而且跟进研讨会最好在没有干扰的中性场所进行。

重要的是合伙人制中的参与者要同意按照一套大家都已经同意的标准来评判该项目的

业绩表现。可文等(Cowan et al)提出了一套简单而正式打分的调查问卷，其中除了成本控制、安全记录和时间安排等标准以外，还包括了对于合伙人制关系的标准，如团队合作和解决问题等。将合伙人制的概念引入澳大利亚的新南威尔士州建筑业生产率皇家委员会，于1992年在其试验性研究中使用了10分制合伙人制效能监控表。表4.3 显示的是一个典型的合伙人制评估概要，或者说是合伙人制效能监控表。

表 4.3 合伙人制评估概要

合伙人制评估概要								
编号	目 标	上一阶段			本阶段			备注
		重要性	级别	分数	重要性	级别	分数	
1	质量达标	15	4	60	15	4	60	
2	造价控制	15	4	60	15	5	75	
3	时间表现	15	3	45	15	4	60	
4	团队合作	10	4	40	10	4	40	
5	安全	10	3	30	10	3	30	
6	避免工业纠纷	10	4	40	10	2	20	
7	避免诉讼	5	4	40	5	4	40	
8	满意的资金周转	5	5	25	5	4	20	
9	对环境的影响	5	3	15	5	3	15	
10	高涨的士气和工作满意度	5	4	20	5	5	25	
合伙人制中各干系人的全面满意度		100	1~5	375	100	1~5	385	

对于工期长于一年的工程建设而言，每隔1~3个月要有一份正式书面的合伙人制评估报告，而且每3~6个月举行一次跟进团体会议。这些正式的评估过程都将成为跟进研讨会议程的重要组成部分。

除计划以内的跟进研讨会外，还会召开一些非正式的合伙人制会议。虽然这些非正式会议对于保持合伙人之间的良好关系十分重要，但如果当新成员加入合伙人制团队时，由正式的研讨会，而非非正式会议来介绍他们也是同样的重要。因为存在着合伙人制的精神在其被传播的过程中有被冲淡的危险(特别是对新的合伙人来说)。

执行阶段的合伙人制的陷阱

项目合伙人制是一种高风险，高回报的方法。很多在合伙人制实施过程中出现的问题也会很自然地在传统的非合伙人制的项目采购过程中出现。然而，虽然与传统的项目采购方式相比，各干系人(那些参与合伙人制过程的各干系人)在合伙人制过程中可能获得更大的利润，相应的，如果走错一步，他们也会面临遭受更大损失的风险。

古博尔(Kuball)列举了六项标准作为合伙人制成功的基本要素。也可以说这些标准对于非合伙人制的项目采购的成功也同样重要。但是，如果下边概括的这些标准不被采用的话，合伙人制过程将面临更大的风险。

- 小组成员之间不能有不牢固的连接。他们必须都能够胜任和完成自己的任务。(奔尔特(Bennett)和杰思(Jayes))进一步建议，小组成员可以允许只犯两次"错误"，之后就要

考虑替换成员了)
- 业主在项目中必须得到适当的体现，并且积极主动。
- 合伙人小组的组成在项目过程中应尽可能地保持稳定。
- 合伙人制应该在设计阶段就开始。
- 有必要任命一个小组领导人，以保证合伙人制原则不会得不到重视。
- 主要分包商和材料设备生产厂商的参与至关重要。

这六项标准强调指出了合伙人制关系中的弱点，同时指明了成功实施一项合伙人制方案所遇到的各项挑战。"内部组织联盟的缺陷源于一组普通的炸弹——对关系构成威胁的弱点。合伙企业是一个比独立企业更为灵活的动态实体，这主要是由于其在组建过程中体现出利益的复杂多样性。一个合伙人制建立了；影响它的各项参数(因素)不会在一开始就完全清晰的显现出来，各合伙人也不会完全承诺，直到之间的信任被建立起来。然而信任需要时间去培养和发展。当各种事情随时间被展开时，合伙人才开始意识到他们所参与的合伙人制内部的分歧和影响。"

当然没有具体的证据证明合伙人制会永远有利于建设项目的采购。虽然在一方面合伙人制可以被看作是"既节约钱又使工作充满趣味性"的方法，但它仍存在着缺陷。成本就是其中一例。合伙人制的直接成本包括研讨会成本以及合伙人制促进员的费用、会议地点租用、交通费等等成本。各合伙人参加研讨会等的时间不是附加的合伙人制的直接成本，而是通常或至少应在传统项目采购过程中也会出现的成本的再分配。另外，下面列出的是一些其他潜在的缺陷和非直接成本：

- 陈腐观念——这是由于当同样的合伙人处于稳定的关系中而失去寻求新合伙人的刺激，因而缺少激励。
- 咨询的费用可能会比较高——合伙人制过程产生出更多可选方案。因而设计咨询公司的人员就会有更大的工作负荷，因而其咨询费也应该更高。
- 减少职业前途——参与合伙人制的职员会认为这是额外的工作负担，而且对他们的职业提升将不会起任何作用。
- 丧失机密——除非采取合适的安全保卫措施，否则与此相关的问题就会发生。
- 投资风险——当单个项目合伙人制不能延续成长期合同时，对联合项目的投资，比如电子数据交换系统的开发可能是有风险的。
- 依靠风险——合伙人变得太相互依赖也是个风险。因此他们应该将合伙人制合约扩大至一系列合伙人。然而在有些情况下，为了保证供应商的继续存在，对单一来源供应商的短期财政帮助是有必要的。
- 腐败——非合伙人制项目的只一次性的关系减小了腐败的可能性和机会。在也许会变成长期性的合伙人制中应该有附加责任，以便保证不会松懈对腐败的检查。

与合伙人制的所有这些潜在缺陷有关且值得警惕注意的是，为了成功，合伙人制需要获得所有合伙人同等级别的承诺。承诺的不均衡性经常是从团体之间的基本差异上发展出来的。例如，一个小承包商在加入到由大型政府机构管理的公共建设项目过程中，可能会感到将时间、人力和物力投入到互相平等的合伙人制中是不可能的。为了平衡双方的承诺，需要所有合伙人的共同努力。

如果这个承诺已经具备，那么可以避免合伙人制实施过程中的许多陷阱。

合伙人制的局限性

评论员都有一个普遍的共识,即合伙人制的参与者不应把他们所有的鸡蛋都放在一个篮子里,而且应该限制他们对合伙人制承诺的范围。早在1981年,英国国家经济发展部建议:"总体来说,任何一个政府的合伙人制项目的实施都不应使用超过其所有资源的30%,而且一个承包商对于合伙人制的全部承诺不应使用超过其全部技术和管理资源的50%。"

完成阶段

合伙人制过程的完成阶段尤其重要,这也是合伙人制与传统项目团队方法相比较所特有的优点。传统的项目团队方法几乎没有一个正式的程序,来总结概括在工程项目过程中积累下来的知识。

项目合伙人制过程的完成阶段通常以最后一个专题讨论会的形式展开。一般来说,这个专题讨论通常会包括午餐或晚餐,高层管理人员将借此机会表扬已完成项目的特殊贡献和强调该项目的显著特征。虽然这种社会活动为人们抒发情感、畅所欲言提供了机会(尤其是如果项目进展顺利),但这个最后的研讨会不仅仅是为其歌功颂德的。这是一个要进行一些认真的工作来从这个已经竣工的项目中汲取经验的大好机会。首次研讨会和跟进研讨会要努力建立起团队精神并促进手中项目的实施和业绩表现,这个最后的研讨会则致力于巩固合伙人制中不同团队成员间的继续合伙人制的合作关系的基础。同时这也是重新采用初始研讨会的优势-弱点-良机-威胁(SWOT)分析法进行分析的良机。最后研讨会的参与者应该从项目成果和合伙人制过程两方面进行有目的的分析,看看哪些方面进展顺利,哪些方面不够成功。如果正式的评估程序(见执行阶段)已在项目进行期间被切实执行,那么项目的进展情况应该已被全面记录下来(缺陷及全部),而且这将为最后研讨会提供一个很好的有目标性会议议题。

如果最后研讨会参与者的数目不会过多的话,在理想的情况下,该研讨会的参与者应包括所有首次研讨会的参与者以及项目进行期间加入团队的关键的干系人。应该认识到的是,最后研讨会是集中工作的重要时机,并且可能成为从单个项目合伙人制过渡到策略合伙人制的重要一步。为达到此目的,最为重要的是委托方应像在首次研讨会时一样全面参加最后研讨会。

策略或多项目合伙人制

已有明显证据表明,策略或多项目合伙人制将比单一项目合伙人制会带来更多的好处。策略或多项目合伙人制与单一项目合伙人制的不同之处在于策略或多项目合伙人制有一个额外内容,那就是发展一个致力于长期效果的,更广大的工作框架。

策略或多项目合伙人制通常出现在合伙人制团队要一起开展一系列项目,不管是新建项目还是改造项目,而且象单一项目合伙人制一样,可以在传统的项目采购方式范围内进行操作的位置上进行。虽然策略合伙人制可以带来显著的利益,但在其执行过程中也存在着需要克服的障碍,尤其是考虑到与自由贸易有关的国家法律和国际法律。例如在美国,部分由于不信任法律的制定的原因,超过90%的合伙人制项目都采用单一项目合伙人制,与此同时，英国财政部采购中心预言在接下来的几年中，合伙人制将会有显著的增长，在其声明中这样暗示道："政府政策是利用竞争来获得最佳效益并促进供应商的竞争能力。但是政府也同样赞赏合伙人制所带来的利益，并且在适当的情况下，这些合伙人制的安排还会得到积极的促进。"

虽然自由贸易对合伙人制特别是策略合伙人制有所限制，可以说策略合伙人制拥有单一项目合伙人制的所有优点，甚至更多。策略合伙人制的长期合作的特点是允许互惠互利的物质基础的发展，例如，共用办公空间和电子数据交换。虽然这些做法对于单一项目合伙人制同样可行，但是对于高科技、高成本的发展计划，如电子数据交换而言，策略合伙人制的半永久性的关系则可带来更为显著的经济和技术利益。

策略合伙人制的参与风险似乎不比单一项目合伙人制的高，虽然多数情况下，但不是总是如此，策略合伙人制的资金周转额度和工作量要明显高出单一项目合伙人制。值得注意的是要使任何合伙人制顺利开展工作，合伙人必须将各自过硬而且特有的技术带进合伙人制中。如果各位合伙人所带进合伙人制中的技术和优势是相同的或者相似的话，合伙人制就变得没有意义了。同时，任何一位合伙人也不应该有慢慢的将其他合伙人吞没的念头。这些要点对两种合伙人制均适用，但尤其适合于策略合伙人制关系。

大多数策略合伙人制将从单一项目合伙人制中所获得的经验演变而成。因此策略合伙人制中的干系人可能在合伙人制机制方面有丰富经验。很可能策略合伙人制的各合伙人已经在一些项目中有过合作。策略合伙人制进行的是一种宏观操作而单一项目合伙人制则是在微观层面进行操作。库波尔（Kubal）用"第二层次合伙人制"这一术语来描述长期或者策略合伙人制，并通过从制造业选取的实例来论证"第二层次合伙人制"这一概念。这里，在制造商和材料及设备供应商之间已经确立的合伙人制计划采用了"准时制"的材料及设备登记系统，因此达到使供应商获得稳定的工作量，制造商削减成本，顾客获得物美价廉的产品的目的。

虽然策略合伙人制的目的与项目合伙人制的目的相类似，他们是不同的。虽然诸如时间、成本、高质量服务以及价值等术语在项目和策略合伙人制中被同等使用，但是它们的使用环境和时间比例是不同的。例如，虽然策略合伙人制的机制看来似乎与单一合伙人制的较为类似，也同样包括首次研讨会和跟进研讨会以及使用外部促进员，但是这些研讨会的时间安排却非常不同，比如即从合伙人的选择到策略合伙人制关系下的第一个项目的启动，期间可能需要有 6 个月的时间。另外，从决定采用策略合伙人制之时起到第一个项目的正式启动之间还会消磨掉 6 到 18 个月的时间。

策略合伙人制的目标应该在合伙人制章程的签署过程中得以体现，这与单一项目合伙人制章程类似。但是项目合伙人制章程签署人的数目可能会达到 25 人左右，策略合伙人制章程的签署者则会被限制在每个合伙人有一个代表签名的范围内。章程会反映出各合伙人的价值观、信仰、哲学观和文化背景，而且这些内容通常会在章程中通过使命宣言和一系列的目标表现出来。下面是一个典型的策略合伙人制章程的模式（图 4.4）：

从合伙人制章程中可以看出，它的目标定得较为广泛而基本，并未涉及到具体项目。有些章程可能包括这样的声明："本声明中不包括合伙人制或者相互约束的协议"以强调协议的非法律性本质。由于此章程将在不同的办公室和工地张贴，人们会觉得对章程的非法律性本质的强调是多余的，因为这种强调贬低了合伙人制章程中的相互信任的精神。

```
┌─────────────────────────────────────────────────────────┐
│                    策略合伙人制章程                      │
│                                                         │
│                      使命宣言                            │
│  我们的使命是为不断提高企业团队间的互相信任的精神和商业  │
│  合作的关系而工作，以期为我们的最终用户创造一个良好环境。│
│                                                         │
│                        目标                              │
│         为了我们的建筑使用者，确保物有所值               │
│         在我们所有的关系及合作行动中一种信任感           │
│         证明对我于所有干系人生产力的真实关系注和利益     │
│         建造优秀的楼宇                                   │
│         按时间在预算范围内建造出没有缺陷楼宇             │
│         为所有干系人，确保合理的盈利                     │
│              公司 A _____                     │
│              公司 B _____                     │
│              公司 C _____                     │
│              公司 D _____                     │
└─────────────────────────────────────────────────────────┘
```

图 4.4　策略合伙人制章程

合伙人制在法律和合同上的意义

[注：合伙人制在法律和合同上的意义将根据当地法律系统随着国家间的不同而有所差异。下面的评论主要针对有着共同的法律系统作为基础的英国、澳大利亚和新西兰三国而言，但即便如此，从各国贸易实际操作角度而言，在此三国中间依然存在着明显的差异。欧洲的竞争法律这一附加因素也对英国产生影响。]

直截了当地说合伙人制章程未在合伙人之间建立起法律关系，但这并不意味着什么都不存在。当然期间的关系力度当然远比真实合伙人制中的薄弱。后者中每个合伙人都被认为是合伙人制的代理，因此可以对第三方(信托责任方)产生约束。然而像澳大利亚建筑业协会(Construction Industry Institute Australia, CIIA)对于合伙人制的调查报告中提到的：

"虽然建设合同提供了权利与义务的框架，合伙人制仍会对风险的分担构成影响，这些风险来自于合同及辅助合同。如果合伙人制的安排出现问题，合伙人会发现其自身正处于一个必要的，或者至少是引人注目的位置，可以宣称，合同的风险分担已被合伙人制章程中的条例，或者是被合伙人制过程中引发的行为表现所改变。这可能就是澳大利亚的合伙人制的主要风险所在。"

在本章中，我们并不想继续讨论合伙人制的法律细节，因为它是随着国家的不同而不同。但是值得强调的是，不论是策略合伙人制还是单一项目合伙人制，都会在所实施的国家中产生影响。下面例子摘自于澳大利亚建筑业协会(CIIA)的报告，它从以下几方面描述了合伙人制可能对建设合同的影响：

- 含有良好信义的合同责任的意义
- 信托职责的产生
- 误导和欺骗行为

- 允诺阻止和自动弃权
- 秘密性和无偏见性讨论

关于良好信义，澳大利亚建筑业协会（CIIA）报告给出以下意见："尽管澳大利亚法院还没有准备赞成以良好信义履行合同作为一种责任，但是有这样一个阐述，当合伙人制的安排出现问题时，应该让当事人公开讨论。因为合伙人制的安排的基本特征是与良好信义相一致的，而且是一个由法院执行仲裁时所应用的良好信义作为支持的美国的概念，那么实施良好信义作为一个普遍责任被引用在合同关系中。因此，合伙人制会导致这样一种局面，即法院准备认可良好信义这一概念；因而当事人的严格实施其合同权限的权利可能被实施良好信义之责任所限制。"应该指出的是，良好信义并不等同于做事合情合理。它是一个广泛性的主观概念。合情合理从法律用语上来看是一个范围较窄的客观标准。报告继续阐述道："最后，值得讨论的是即使在建设合同中没有提到合伙人制，合伙人制章程也可以形成合同文件的部分内容，因此，章程提供了在合同实施过程中确保良好信誉的明确的责任。"

这种法制观点表明了合伙人制的潜在的合同及法律风险。（读者可以在澳大利亚建筑业协会（CIIA）的报告《合伙人制，成功的模式，第 8 号研究报告》中获得更加详细的有关资料。该报告是专门针对澳大利亚建筑业采用合伙人制的状况。）我们并不认为合伙人制在法律方面充满漏洞。但是试图在合伙人制协议中，将法律及合同双方面的参与风险减到最低限度，确实是明智之举，这可以通过在合伙人制过程的初始阶段对潜在的法律及合同问题进行预测和预防来完成。澳大利亚建筑业协会（CIIA）建议合伙人制的合伙人应该将有关规定及实施步骤统一写进合伙人制章程中：

- 良好信誉——通过在合伙人制的安排中的明确定义来使得这个问题不会被带入合同条款中去。
- 受托人职责——（达到比简单的商业关系更高的行为标准，一个真实的合伙人制的特征）通过明确的规定来排除受托人职责产生的可能性，由此允许合伙人自由发展他们的兴趣；或者采用另外的方法，即控制职责在合伙人制过程的目的的范围之内，由此允许合伙人在其职责范围以外发展自身兴趣。
- 贸易实践法律条款——很难通过所签订的协议来否认误导行为的责任。
- 承诺阻止和自动放弃——（这一内容与未来行动有关，需要加强说明和承诺的力度，包括承诺不依靠合伙人的严格的法律权利，特别是当其可能被过度使用时。）将一个程序并入到合伙人制章程中，如果为与建设合同中的术语和情况取得一致，合伙团体可能会否定其权利以坚持合同的执行，那么此程序势必要被遵守执行。
- 机密性和与机密性有关的"无"偏见讨论——包括一项机密性条款以保护机密信息，通过禁止向除参与者以外的其他人透露信息而达到此目的。关于"无"偏见讨论，明确合伙人制过程中的任何泄露或妥协都是为了达到此目的，包括任何正式文本。

澳大利亚建筑业协会的报告建议这些条款应被编入合伙人制章程中。这样做似乎比较困难，因为章程是一个关于信任和合作的明确声明，而且这个章程会在不同地方张贴。对合伙人制过程而言，在合伙人制章程中出现一套小幅印刷的法律条文是不合适的。然而澳大利亚建筑业协会报告给出了值得注意的良好建议。解决问题的方法也许是将该报告中罗列出的问题整理，不是在合伙人制章程中，而是作为章程的补充条款。澳大利亚建筑业协会报告中提的建议并不意味着又回到传统的对抗的环境中去，而是一种减少法律和合同方面带给合伙人的风险的合理的防范。

那些在英国进行合伙人制安排的公司还需要留意欧洲法律，因为该法律对欧盟境内的贸易产生影响。《相信团队：建筑业合伙人制的最佳实践指南》一文中对于对英国产生影响的合同和法律问题进行了详细的说明。对英国合伙人制有兴趣的读者可以参考这项报告。下文从该报告《相信团队》一文中摘录出来，它清楚的阐释了有关的合同和法律问题相协调的重要性：

"如果合伙人制的影响会导致对本国企业的歧视，或者破坏了欧盟的基础自由——也就是商品、服务、劳力和资本运动的自由，那么该合伙人制的章程就在欧盟法律中是非法的。即使合伙人制的安排并未影响到成员国间的贸易，如果合伙人制章程协议对价格和商品供应施加了限制，它仍可能是无效的，而且要受到罚款的，或者对英国法律认定下的损坏进行赔偿的处罚"。

争议解决

如果假设没有争议，那么就象是乌托邦式的虚幻理想，甚至处于无对抗性的合伙人制关系的文化中，对于有关合同和法律方面的问题，仍需谨慎小心，以防意外不测之事发生。据最近对澳大利亚合伙人制项目的统计，91%被访者同意争议解决方案是合伙人制安排中的一个重要因素。在合伙人制安排没有成功的项目中，43%没有任何正式争议解决方案。

虽然在合伙人制安排包括争议解决问题的程序很重要，但仍须承认合伙人制关系中的争议事件可能相对较少，这是因为合伙人制的实施计划应采纳一种阻止差异发展成为争议的机制。

以下是一个典型的解决问题的过程及方法：
- 在最低一级的权限内解决问题
- 在导致项目延期之前，那些最低一级权限不能解决的问题将由当事人双方及时提交给上一级进行处理
- 越级权限是不允许的
- 忽视问题或者不对问题作出决策是不可接受的

"合伙人制要求提供尽可能多的交流机会。传统方式上，柔软的天鹅绒窗帘经常被项目工作组成员用来掩盖问题并且希望它们能够自己解决……通过一个解决问题的过程，各干系人的好与不好的经验都提到台面上。可以公开讨论合同中潜在的风险和困难。一种高度信任的文化环境形成，在这样的环境中每个人都可以自由发表意见并为问题的解决出谋划策。"

结　论

在英国、澳大利亚和新西兰的建筑业中，也许有许多人对于从美国发展起来的管理概念都持有一定程度的怀疑态度，正如合伙人制的例子，尤其是如果对该概念进行大肆宣传的情况下。对于一些人而言，对双赢商业环境的承诺似乎是一个不可能实现的虚幻的梦。但是确有许多记录在案的成功使用项目和策略合伙人制的实例。奔尔特（Bennett）和杰斯

(Jayes)做出了如下评论:"对研讨会的最初投资和对于合伙公司的慎重选择会很快转化成显著的净收益,对单个项目合伙人制而言收益可以达到总成本的10%。随着时间的推移,策略合伙人制可以达到30%的成本节约的目标。"

在澳大利亚,澳大利亚建筑业协会关于合伙人制的调查发现"将近85%的被调查对象会继续采用合伙人制项目。"

不会令人吃惊的是,那些参与过合伙人制项目而又没有合同纠纷的人们认为,合伙人制可以是巨大的成功。相反的,当合同纠纷超过5%时,合伙人制被认为是失败的。在成功的合伙人制的实例中,有56%在继续合伙人制的合作。而在那些失败的例子中,再没有继续合伙人制的合作。值得指出,从澳大利亚建筑业协会的调查发现,虽然合伙人制被调查对象看作是成功的,并且那些经历过成功合伙人制的人们会再次成为合伙人,在一个以五分制来衡量合伙人制的成功和热情的调查中,平均3.83分是"非压倒性的高"。

因为合伙人制与人际关系和信任有关,它不会是永远没有错误的。然而有关证据表明合伙人制作为一个建设管理概念是受欢迎的和很有利的,这一概念能促使建筑业的文化从对抗性转为非对抗性,因此它将明显提高所有干系人的净收益。合伙人制的主要优点包括:

- 减少法律讼争
- 项目的成本、时间和质量都取得了提高
- 较低的管理和法律费用
- 提高了创新和价值工程的应用机会
- 提高了财政投资成功的机率

一系列更加详细的优点在澳大利亚建筑业协会的研究报告中得以体现,并按其重要性排列如下:

- 专业知识的相互交流
- 减少法律讼争
- 较低的行政管理费用
- 因为双赢的态度使财政投资得以成功
- 对索赔所带来的成本有积极作用
- 对施工进度工期等有积极作用
- 更好的时间调控
- 更高质量的产品
- 促进新技术应用
- 更好的造价控制
- 鼓励创新
- 减少重复工作
- 提高安全作业
- 更大收益的工作
- 图纸等文件的错误更少
- 来自于信息交流的创新
- 记录创新的机制

在本章的介绍部分，我们以这样的评语作为开头："合伙人制不仅仅拘泥于旧时代的价值模式，或者回归到'君子无戏言'的美好的旧时代。"合伙人制若要获得成功，需要通过计划和实施方法来实施。我们已经描述过的理想的合伙人制"在这方面关键表现指标起着重要作用。对工作表现的连续监督是合伙人制获得成功的最关键的因素之一"。

除却需要保持良好的各方合伙人的交流，将交流情况记录下来以使各合伙人对项目状况的发展有个全面了解也是十分重要的。但是合伙人制不只是高级理想和创造相互信任的环境，它还包括适当拥有良好的组织系统以及强调按部就班地完成各项任务，诸如保存适当的记录。例如澳大利亚建筑业协会研究中被调查者认为合伙人制的会议记录是不断对该项目进行评估的最重要的特征和依据。

合伙人制的概念好比奥运会的火炬，当火炬被接替传递下去的时候，存在着火炬掉落，火焰熄灭的危险。同样的，如果合伙人制的理想不进行持续地修订，精心地培养，那么合伙人制的理想很可能在传播的过程中被歪曲。

第五章

价值管理(Value Management)

导　言

多数人都赞同成本与价值之间存在着差异，并且价值较成本更难定义。在拥有卫星电视和复杂的音响设备的现代家庭中，用电池作为操作能源的收音机的价值是很小的。但是当暴风雪切断了电力供应，而由电池启动的小收音机便成了收听新闻的惟一工具时，其价值得到了彻底的提高。

在英语字典中，有几个关于价值的定义，从"使任何东西有用或者可估价的"到更为简单的表述"价格"。正因为价值一词中包括了这些不同的概念，价值管理有时会显得难以理解，并且可能因为此原因，使得有人对价值一词与成本节约、可建设性和成本计划等概念相混淆。这样的混淆自然导致了价值管理在应用上的分裂，以及在当今的建筑业中此项技术还徘徊在其发展道路上的十字路口上。继 20 世纪 80 年代末期的一阵狂热之后，对于此技术的使用和兴趣大幅度降低，这主要因为缺乏对于价值的不同概念的正确评价。这在价值管理的历史发展进程中得到了清晰的体现，其历史发展可以从现今在建筑业中的应用追溯到 20 世纪 40 年代时的创建。

发展历史

价值分析这个名词首次是由麦尔司(Miles)用来描述他在第二次世界大战期间在通用电器公司发展的一项技术。这项技术开始时是为了寻求可替换的产品零部件，以弥补第二次世界大战所造成的短缺。然而由于战争的原因，连这些替代的产品零部件也经常无法得到。因此这导致了不再只寻求可替换的产品零部件，而是寻求一个可实现该零部件功能的另一种方法。后来人们发现这个功能分析的过程能够生产出更便宜的产品，而且没有降低产品质量。大战之后，这套系统被保持下来，并用于减少产品不必要的成本和改进设计。

麦尔司(Miles)理论的中心特征是，对顾客需要的产品功能的定义。这些功能只用一个动词或一个名词来定义。麦尔司(Miles)相信，如果不能做出如此定义，那么产品的真正功能就没有被理解。继而以可能的最低的成本来实现功能的方法对该功能进行评估，而且这个评估被用来作为找到能实现该功能的不同方法的依据。麦尔司(Miles)以电子发动机屏幕为例来说明他的做法，此电子发动机屏幕需要具备以下功能如，如表 5.1 所示：

接下来是通过选择最低廉的可行的实施方法来评价这些功能。

- "排除物体"的功能是以用于覆盖发动机的一块金属薄板的成本作为评价基础的。
- "允许通风"的功能是建立于在金属薄板上挖洞的额外成本上。
- "促进维修"是通过以便于拆除该金属薄板而增加的一个弹簧夹的成本来进行评价的。
- "使消费者满意"是以对该金属薄板油漆上颜色的成本为基础的。

表 5.1 功能定义

	动词	名词
1	排除	物体
2	允许	通风
3	促进	维修
4	使满意	消费者

这些成本如表 5.2 所示:

表 5.2 对功能进行成本计价

	动词	名词	实现该功能的最低廉的方法	实现该功能的最低成本
1	排除	物体	金属薄板	$0.15
2	允许	通风	金属薄板上面的洞	$0.15
3	促进	维护	弹簧夹	$0.10
4	使满意	消费者	金属薄板的油漆	$0.10
实现全部功能最低总成本				$0.50

因此所有这些最低成本可以被看作是屏幕的真正价值,因为仅这些成本就可以满足顾客要求的全部功能。通过这一过程,麦尔司(Miles)以获得功能所需的可能的最低成本为基础确定了该电子发动机屏幕的功能价;本例为 $0.50。接着将这个可能的最低成本与目前的实际成本 $4.75 进行比较。这个例子清楚地说明在制造该屏幕的过程中有大部分所付出的成本并没有获得功能。

定义和评价功能的同时也为获得以实现该功能的进一步的方法提供了便利。这一点可以用一个建设项目为例进行说明。在一个酒店建设项目里,建筑师将一个儿童戏水池设置在酒店中心游泳池旁边。该戏水池的作用便不可能像原来设想的一样——为儿童提供休闲嬉戏的场所,而事实上相当于一套安全设施将他们隔离在中心游泳池以外。一旦儿童戏水池的功能被定义在"保证儿童安全"上,则可以更为容易的找出多种选择方法。一个游戏区,一个小型游乐场或者甚至一个托儿所都可以满足这些要求。最后设计组将喷泉定为最终方案,其功能既满足了安全的要求,成本又比最初的戏水池低许多。

这种功能定义,功能评价和多种选择性方案的产生被统称为功能分析并且它是形成价值工程基础的基本方法。

很明显,在复杂的企业组织中,需要一个可以进行功能分析的系统:功能分析不能在特定的基础上进行。当要进行功能分析时,要考虑这些问题:由谁去做?如何组织?因此,价值工程系统,像它如今被称呼的名称一样,围绕着功能分析技术而发展起来。图 5.1 表示了一个发展完善的典型的价值工程系统。系统中心是功能分析的概念。研究分几步完成,即工作计划,并且在产品生命周期的某些时候由一个价值工程小组来实施。

直到 20 世纪 70 年代价值工程仅被用于制造行业,然而在同一时期它也被美国应用于建设项目。主要的应用都是政府有关部门,他们由于各种原因采取制造业价值工程系统,并加以改造,以适用于他们各自的使用目的。

第五章 价值管理(Value Management)

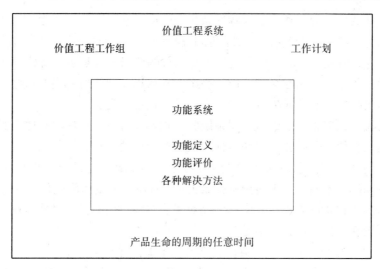

图 5.1 以功能分析相关的价值工程系统

如图 5.2 所示，当价值工程开始被应用于建筑业时，40h 的研讨会便成为了价值工程的一个特征。另外价值工程在设计完成了 35% 的时候实施，且使用项目团队以外的团队，即非设计组本身来完成这项工作。这个价值工程系统很大程度上源自美国，而且在英国建筑业中被首次使用。

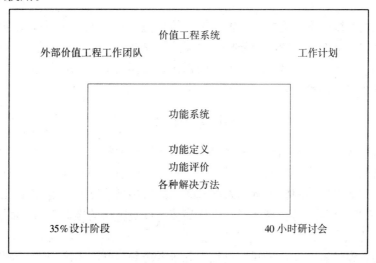

图 5.2 建筑业中的价值工程系统

这个在美国仍然被广泛应用的系统存在的问题是，它偏离了麦尔司(Miles)原始理论的方向。除了在名字上，它没有包含功能分析。价值工程在美国基本上就是设计审查。它只是一个松散地围绕工作计划组织的 40h 研讨会。并由一个外来小组在设计完成了 35% 的时候进行。该价值工程小组的工作主要针对项目中高成本的分项并对此找出不同的解决方案。挑选高成本分项是一个相当松散的过程。它基于高成本构件成本与其他较便宜的方法的成本的比较，以及对项目造价的一般分析。这种模糊的方法导致了价值工程的广泛的和伴随着各个专业的设计变更和成本削减的结果。然而这些结果不能作为是功能分析。作为一个独立的内容，40h 的研讨会本身对价值工程研究的成功是一个关键的有贡献的因素。在 40h 研讨会内，研究讨论的成功程度与参与人员的性格尤其是领导者的性格、研究讨论的时间、价值工程小组的相互交流、设计小组的介入以及业主所充当的角色，都有很大联系。功能分析的技术与研究成果有很少联系，甚至根本没有关系。

当英国公司尝试应用美国价值工程系统时，他们发现并不能使之有效。其中有许多原因，但主要是与价值工程研究的最初目标有关。美国价值工程系统是因为政府建设项目需要更强的可说明性和责任以及义务而产生的。(几乎美国所有价值工程行为都是政府项目。)而英国的情况就不同了。工料测量(quantity surveying)系统已经提供了所需的说明、责任和义务。价值工程是用来提供一个检验相对于成本的价值的平台。从这点上来看，美国的价值工程系统在英国失败也就不足为奇了。

当这个失败在英国发生以后，英国建筑行业面临两种选择。其一是干脆废除价值工程；其二是回归到麦尔司(Miees)的起始理论，建立一套能满足其原始目标的系统。他们选择了后者，而且发展了名为价值管理的新系统。因此，此文从此之后只使用价值管理这个名词。这个选择意味着，实质上英国是在一张白纸上开始建立新的价值管理系统。然而如上面所提及的，美国的经验也表明，在价值管理系统中，除了功能分析之外还有其他内容，例如价值管理研究小组的组成和研究时间的长短也同样会影响价值管理的成功与否。为了建立一个有效的价值管理系统，这些内容都需要调查研究，本章的下一部分会详细讨论这些内容。

然而在我们继续往下讨论之前，关于美国的价值工程系统的最后一点需要澄清一下。作者将价值管理定义为综合的功能分析。然而这只是我们的定义。钱伯斯(Chambers)字典所解释的价值一词包含了许多意思。美国的系统，虽然忽略了功能分析，但仍然很有用，并且实现了它所要求的目标。

功能分析

在我最近一次去新开的当地一家诊疗室时听到护士抱怨该建筑设施上的不足。在我们候诊的那间房间里，门直接朝向工作间开启，这意味着病人的活动病床不可能被直接推进或推出房间。因而需要使用轮椅将病人推进病房然后抬到病床上。她(该护士)抱怨道如果将门开在另一边，就会变得容易多了，这样病床可以被自由的推进推出，而不被工作间阻隔。

很明显的，设计者在设计房间的过程中并未考虑到房间的未来用途。他没有想到此房间会被用作为那些行动不便、无法自行步入房间上床的患者诊病的检查室。也就是说，他没有恰当的定义该房间的功能。其中最可能的原因是他没有在设计的过程中让房间未来使用者参与讨论。那么该房间的价值因为它未具备所要求的功能而被大大削弱了。

建筑或建筑的某部分未达到预期的功能要求是建筑业中出现的普遍问题，因此这也成了功能分析的基础。在功能分析的过程中，功能的要求与价值获取之间有着密切的联系。当所有功能都以最低的成本得以实现时也就取得了良好的价值。如果功能无法实现，或者以高成本获取这些功能，则意味着只具备了很少的或不具备价值。

然而问题在于即使设计师能够定义出特定项目所具备的功能，他如何知道这些功能确实能以最低成本获得呢？正如前面已解释过的，在功能定义之后所进行的功能评估，是对以可达到最低成本为基础的定义的功能进行评价。例如，设想有一扇窗户，其具体的位置也许使其只具备采光的单一功能。实现此功能的可能的最低成本就是在墙上开个洞。因此该行为的成本就是此功能的价值。当然这个例子不能只照字面意义理解。绝大多数情况下，窗户的功能不可能只是采光。这个例子是为说明任何为获得功能以外所付出的成本都是不必要的并且降低了其相应的价值。在许多炎热地区确实有采光的功能是通过在墙上挖洞实现的实际例子。

功能评估的另一优势是它具备创造性。它可以帮助发现其他一些同样可以满足功能的解决方法。在寻找可以实现其功能的可能的最低成本的过程中引发了这样的问题："如果墙上的洞可以满足采光的功能，那么还有哪些方式可以达到同样的目的？"屋顶上的洞口，透明的墙体，或者电灯都可以具备采光的功能。

正如图 5.3 中总结的一样，功能定义，功能评估和具有创造性的可选择的方案这三项被统称为功能分析，这正是价值管理的基础核心。如果没有这项技术，项目中实施的任何行为，虽然也许有用，却不能算作价值管理。由于这个原因，这一方法中每一阶段都要进行进一步的研究。

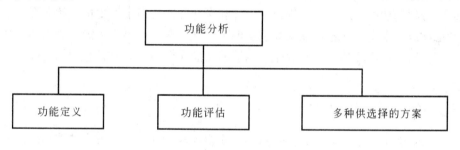

图 5.3　功能分析阶段

功能定义

功能是通过一个动词和一个名词来定义的，因为这样简单明了。一般来说，如果一样东西的功能不能用一个动词和一个名词来定义的话，那么它的功能还未被完全理解。

建筑业中伴随功能定义所产生的问题是层次问题。在上述例子中，一个建筑构件，也就是窗户的功能已被定义，但是窗户也可被划分为一系列的组成部分，如窗头、窗台或者窗锁，这些部件均可被定义。同样，楼宇所提供的空间的功能也可被定义。例如，在一所学校里，这可能包括教室和员工办公室的功能。或者退后一步将学校功能作为一个整体加以定义。在这四个不同的等级上定义功能会产生四种不同的结果。那么应选择这些层次中的哪一个呢？

将项目功能定义为一个整体

很多项目并不像它们看上去那样。例如试想一座桥的功能，其最明显的功能是"交通功能"。然而也许并不一定是这样，其真实的功能也许是将工业基地转移到城市的另一地区，或者为建筑业提供工作，或者减少交通阻塞。在这个层次上的正确的项目功能的定义是非常有用的,因为如果该桥梁的正确的功能是将工业基地转移到城市的另一地区,那么很明显,除了建设一座桥之外,也可以通过其他方式来实现此目的。给城市的另一地区提供经费就是另一种方法。

定义项目内空间的功能

一旦确定需要某一项目,项目中的空间功能就可以被定义了。例如在一所学校里,我们可以问教室、操场、员工办公室或运动设施的功能是什么。操场的功能可能是为孩子们提供放松娱乐,呼吸新鲜空气的场地,或者使教师们获得更好的休息。取决于不同功能,设计的方法和任何可行的选择的方案都将不同。

定义分项的功能

除了项目和空间功能,项目分项也可以被定义。再一次以窗户为例,你也许会将窗户的功能定义为"通风"或是"采光"。但是考虑一下监狱窗户的功能。此时,其功能将不是为了通风,因为这扇窗户很少被开启。其功能也不是采光,因为即使在白天,光线通常也是不充足的。在这种情况下,窗户的真正功能是"使环境变得更为人性化"。

因而,分项功能的定义也可以提供对设计和设计方案的清晰的内情。然而在一般情况下,不考虑建筑物本身,就不可能给分项功能下定义。永远不要简单的认为内墙的功能是"分隔空间"。监狱中内墙的功能与卫生间内墙的功能完全不同。而且它们的功能都不是分隔空间。

定义分部功能

将分项分解为分部是可能的,而且这将再次提供对设计和设计方案的清晰的洞察。将一扇窗户分解为分部件,如窗台、窗头和窗框,可以检验这些分部满足所需功能的程度。

应选择哪一层次的功能?

在建设项目中使用的功能分析有四个层次。那么应该选择到哪一个层次呢?对此的答案取决于业主,设计完成程度和项目的总体情况。如果业主只要求得到一个项目的概念,那么对项目本身功能的定义则不一定必要,而对空间功能的定义可能是最为适当的。同样的,如果项目所提供的空间已被限定,那么定义分项功能则可能是提高价值的最好方法。通常在建筑业中不会采用分部功能定义法,因为一般认为它更适用于制造业。

当然可以在同一价值管理研究中将全部三种或任意两种功能进行定义。但是这样做可能非常复杂并且浪费时间,尤其是面对大型建设项目时。图 5.4 概括了功能定义的定位。通用规则是功能定义的层次越高,项目变更能力就越大,因此提高价值的潜力也就越大。通常应该选择功能分析的可能的最高层次。

设计之前与之后的功能

高层面的功能定义对项目更为有利。因此功能定义的理想时间是在设计开始之前;从而最大限度地发挥变更的潜能并提高价值。然而通常情况下是设计已经存在,这种情况下的问题是价值管理团队会试图用这个设计来确定功能。这意味着在确定顾客和使用者所要求的功能之前，他们已经按照设计接受了这些功能，然而，归根到底，功能定义应该根据业主和用户的要求，因而由此产生的功能将不会变化，不管设计存在与否。这看上去也许只是细微的差别，但却很重要。很多对设计的选择都是基于对过去做法的认同。但实际上最好的方法是严格根据业主的需要而进行全新设计。如果这样做的话，设计中所包含的功能只是必要的和需要的功能。

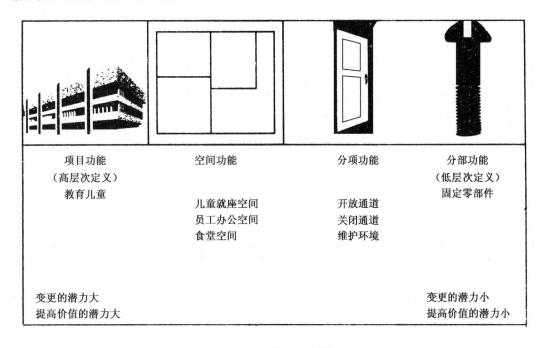

图 5.4 功能定义级别

功能分析系统技术(FAST)图表

美国价值管理人员也认识到了不同层次的功能定义所带来的问题，他们试图通过使用功能分析系统技术也就是 FAST 图表来显示这些层面间的内部联系，如图 5.5 所示。图 5.5 从主要功能或高层面功能开始，并提出这样的问题："如何实现这些功能？"其答案形成了下一个层面的功能，并且继续对其提出同样的问题，直到实现最后的功能。与此相反从图表的最低层面着手并问"为什么"来检查该图表的逻辑性是否正确。图 5.5 中所示的例子是关于防撞护栏的。其最高层次功能是通过尽量减小伤害，引导交通远离危险，在事故中减弱震荡并增加对危险的意识来实现挽救生命的目的。(以下只显示了图表的部分内容)在事故中减少震荡是通过吸收、转移并改变能量的方向来实现的。

所有的美国价值管理教科书中都会包括 FAST 图表的示例，但从我们的角度来看，没有一个可以经受住即便是最温和的学术审查。本书作者认为 FAST 技术，至少在建筑业中，对于那些非常复杂且耗费时间的项目是不起作用的。

成本可否被分配到功能上？

一些价值管理从业者在功能定义结束时将项目的预算评估成本按功能进行分配。例如，如果一家医院的主要功能被定义如下：

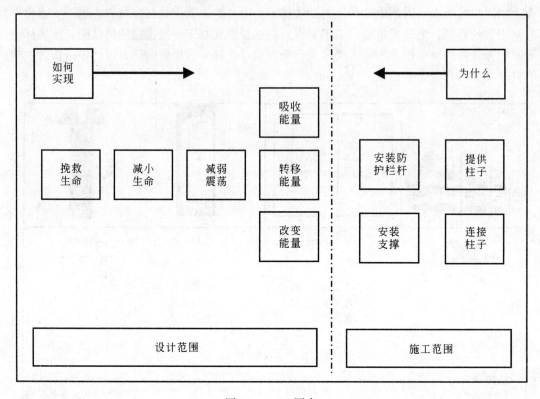

图 5.5 FAST 图表

- 治疗病人
- 诊断病情
- 留院治疗

然后该项目的预算成本需在这些功能间进行分配，以便业主明了每一项主要功能会花掉他多少成本。这种方法被何华德·意利甘(Howard Ellegant)和其他著名的美国价值管理从业者所应用。这种方法有一个优点，那就是业主意识到某一功能可能涉及投资额很大的话，他可能不再想将该功能包括在项目内。然而，把成本分配到功能上是很难的，因为有些分项，例如基础或中央设备，不论有没有包括其功能，都必须具备。另一个将成本分配到功能的问题就好比拿苹果和梨相比较，风马牛不相及。我们已经看到功能定义不应与已有设计相联系，而最好是独立实施。如果是这样的话，功能定义与预算成本之间将没有任何关联，因为后者只是已有设计的反应(图 5.6)。正因为如此，本书作者认为按照功能分配成本是毫无意义的。然而，有对此提出不同意见，也是很好的事。

功能评估

一旦功能被定义，它们将以可实现该功能的最低成本为基础被评估。在前面的例子中我们定义一扇窗户的功能为采光。实现此功能的最低成本是在墙上挖个洞。然而不能直接按字面意义来理解；这只是解释功能价值的一种途径，并且是寻找能满足此功能的可供选择的方案。实际上我们无须对墙上的洞做准确定价，而只是对其成本有一个概念。功能评估的目的就是鼓励有创意的选择方案。而不是表面意义上的对成本的计算。

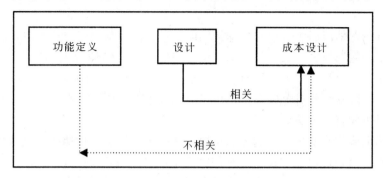

图 5.6 功能定义与项目评估间的关系

因此价值管理研究可能产生两套成本：分配在不同的功能间的项目的预算成本（功能成本）和实现这些功能所需的最低成本。在一些价值管理成本研究中，值得系数是通过功能成本除以实现该功能所需的最低成本计算的，以此来达到用数学表示法表示价值的目的（如图 5.7）。这些系数显示出一些花费在功能上的不必要成本，而且还可以用来对某些优先功能进行再研究。另一种方法就是成本不具体分配到功能上，而将功能看作是实现该功能的可能的最低的成本（图 5.8）。关于功能评估的这两种方法总结一下，作者推荐后一种方法。

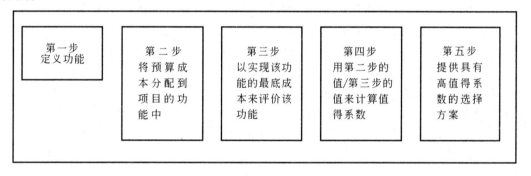

图 5.7 运用预算成本评价功能

图 5.8 不运用预算成本来评价功能

可选择方案

对功能进行评估以后,可以提出这样的问题:"有没有其他的方法可以达到同样功能?"来产生多种不同的选择方案。回到监狱囚室窗户的示例中,"使环境人性化"的最便宜的方法可以是提供一台收音机。但是还有哪些方法也可实现这一目的呢？一台电视,一个流动图书馆,较大的囚室。所有这些方法都可以"使环境人性化",且较之开窗的方法,这些可能是更好的且更便宜的方法。

虽然用最低成本来实现功能的方法可以促使产生多种满足功能的可供选择的方案,但是要每个人在价值管理研究中提交颇有创意并满足这些功能的想法,还颇具难度。因此暴雨式思考的方法经常被用于协助产生多种选择方案。暴雨式思考鼓励野性的令人难以接受的建议,不过我们希望这些建议能促使产生可行的好主意。此举的好处是,因为鼓励和欢迎一些看似荒谬的建议,而没有因此带来难堪,好的主意可能由此而生。在这个时候,最愚蠢的主意也会被看成是最好的。鼓励参与者呼喊出他们想到的任何点子,而不鼓励正式性的发言。(一些作者甚至建议在这个研究阶段,喝点酒作为帮助。)暴雨式思考是一种通用的管理技巧,它不仅仅只针对价值管理。关于暴雨式思考的著作已有很多,但由于本书篇幅的限制,我们这里仅罗列出它的一般规则。

- 尽可能搜集所有的主意,再进行评论。
- 要求有大量的主意。
- 所有的主意都会被记录下来。
- 最佳主意通常出自于经验不多的新手。
- 这是一个团队性的工作,应该在现有主意的基础上增加新的想法和主意。

图 5.9 概括了功能分析,也就是价值管理的核心。它包括三个阶段:功能定义,功能评估和多种方案的产生。功能分析通过一个动词和一个名词来定义；功能评估即对实现该功能的最低成本进行评估，而多种方案则产生于暴雨式思考之中。

功能分析并不容易实施,也许要通过几年的实践才能对其不断完善,仅通过对一本书中几页内容的学习是掌握不了具体方法的。另外它不是一门孤立的技术。为了发展价值管理系统,我们不仅需要功能分析,还需要一套有效的实施办法。如何组织这项研究,由谁负责这项研究及何时开始这项研究都需要仔细的考虑。本章下一节对价值管理系统的这一内容进行阐述。

价值管理研究的组织

价值管理中一个必要的部分就是一个包括五个阶段的工作计划。这五个阶段概括如下:

信息阶段

在这个阶段,搜集项目需要的所有信息。

分析阶段

在这个阶段,功能分析分两步执行:

第五章 价值管理(Value Management)

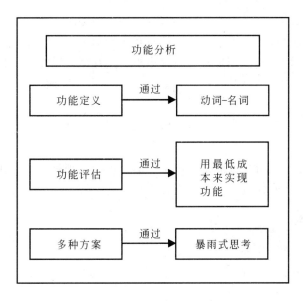

图 5.9 功能分析的概要

定义功能
评估功能

创新阶段

在这个阶段，以功能分析为基础产生能够满足分析阶段所定义的功能的各种可能的方案供选择。

判断阶段

在这个阶段，对各种能满足功能的方案进行评估。

发展阶段

那些被认为值得继续考虑的方案，将在这一阶段被选择出来并被进一步发展。

应该由谁来实施研究？

价值管理的目的是，将对项目感兴趣的人们的需要和想要进行最优化。因而为了做到有成效和成功，价值管理必须以团队形式开展。其实这一点并不是例外，而且从 20 世纪 40 年代至今价值管理就一直被视为团队性方法。价值管理和普通管理领域中的大部分关于团队工作活动和领导能力的研究和文献已被记录在案，但对那些要想进行深入研究的人来说会发现还缺乏足够的信息资料。然而对于那些对价值管理实用性研究有兴趣的人而言，下面的内容是一个概括说明。

团队该由哪些人员组成？

价值管理团队可以由一个对此项目不了解并且没有参与过此项目的设计小组或外部小组来组成。也可以是这两者的混合，有或没有业主参与都可以。另外，如果项目有特殊的问题，或者需要专家的介入，也可邀请一些专家加入该团队。

该由设计小组还是外部小组来实施价值管理会更好这一问题，还有待争论。在美国，较为倾向外部小组，而在英国则更倾向设计小组。这并无对错之分，且最终要由业主做决定。下面是对两种方式优缺点的概括。

采用外部小组的优势

- 目的性强
- 小组成员可根据个人特殊技能挑选出来，而设计小组则是早已存在。
- 对业主负责，再次确保设计小组的设计方案是优秀的。

采用外部小组的缺点

- 设计小组可能很难接受外部小组的出现。
- 设计小组已经组成并克服了很多合作中出现的问题。一个外部小组还要花时间去磨合。
- 外部小组的工作也许并不完全的客观。某些情况下，特别是在面对大业主时，外部小组希望"有出色的表现"，其中一种表现就可能是对于项目的研究过于苛刻；以显示他们可以做得更好。
- 聘用外部小组比聘用设计小组更容易影响设计工作的正常进行。
- 如果功能分析运用得当，那么并不一定需要引入外部小组。
- 外部小组也许会过于考虑削减成本，而没有真正考虑其是否为实现功能价值所必需的。
- 外部小组对项目缺乏深入了解。
- 聘用外部小组价格昂贵。
- 如果外部小组对设计进行了修改，而此修改在后来被证明有缺陷，那么这个外部小组就引起了设计责任的问题。

当然没有十全十美的团队。虽然外部小组的缺点似乎比优点多，但无疑在某些情况下外部小组会有优势；比如当项目在政治上遇到了严重的难题时。然而本书作者的观点是，在绝大多数情况下，设计小组将提供比外部小组更为有效的价值管理研究。

应该由谁来领导团队？

价值管理团队的领导人，或通常被称为促进员，需要有综合的技能包括有深入的功能分析，小组和团队建设和对项目可选择方案进行评估的知识，以及施工知识。设计小组中的任何一个成员都不太可能有这些技能，所以可能有必要聘请一位外部促进员。价值管理是一项以经验为基础的技术，因此通常建议聘请一位外部促进员。

业主是否应该参加研究？

没有业主的出席，就不可能开展价值管理研究，即使是在功能分析的基础阶段。楼宇的未来使用者也可能对于楼宇的功能提供有深度的建议，因此应该邀请他们参加研究，但目前能真正做到这点的还不多。

团队规模和组成

价值管理团队成员的数量不宜过多。至今，还没有人对最优化的价值管理研究团队成员数量进行研究，但是现有的经验告诉我们一个价值管理研究团队的人数最多不能超过8个。

价值管理团队的所有成员都要具有大致相同的资历，否则资历较浅的成员将会感到胁迫性。另外团队成员必须有足够的对项目做出主要决定的权力。

价值管理研究形式

由于早期美国对英国在价值管理发展所起的强烈影响，此方法通常采用40h研讨会的方法。这个系统是将价值管理团队召集到一起达40h，远离通常的工作环境，以便研究工作不受干扰，从而提供一个适于发挥创造力以产生多种选择方案的工作环境。这种研讨会的形式在英国被保留了下来，但很少被认为必须是40h。事实表明，在英国，价值管理团队研究的平均时间是两天。在美国采用40h研讨会的原因主要是由于国防部坚持对价值管理过程实行标准化，而没有其他部门都认为这是最优化方法。美国的40h价值管理研讨会的成本高昂，且价值管理团队必须提供全面的计划文件以及相应的成本清单。根据作者的经验，40h的研讨会中有一半时间被用在成本计算和编写计划上，并将这些文件整理成适合业主阅读的文本形式。在英国，对于成本的重视相对较少，因此该形式被认为不太必要，所以研究可以快速实施。有趣的是美国国防部的价值管理程序对于成本的要求相当严格，以致于在每个研究结束前，都要将新提出的方案可能带来的节省除以该项价值管理研究所花成本，来表示物有所值或者是投资回收率。

由于价值管理是团队活动，它几乎不太可能与研讨会的概念相分离。一些价值管理从业者甚至主张建立一套综合系统；价值管理仅仅作为通常设计过程的一部分来实现。没有证据显示该方法不能实现，但作者认为功能分析技术的目的在于改变思维模式；而如此综合的方法不太可能成功。

应在哪里开展研究？

对于应该在哪里开展价值管理研究，没有严格的规则。但大部分价值管理顾问和促进员都建议离开通常的工作环境，而在酒店或会议中心进行比较好。

研究开始的时间

价值管理研究可以在项目生命周期的任何一点进行。由于项目的功能不会因为设计师在开发过程中的前面或者后面的阶段而改变，所以何时开展价值管理研究不应该对结果产生什么不同。不过随着项目的进展，重新设计或涉及变更的工作也会增加。因此这些设计变更所需的成本也会相应增加，而且设计小组也会不愿再有改变。正像前面提到过的，功能分析可以在不同层次(项目、空间和分项等层面)上实施，而且如果只要求对项目的分项进行功能分析，(可能由于规划许可或者其它限制,空间布局已被限定)那么直到设计达到适当的程度才可能进行功能分析。

一些价值管理作者建议功能分析的层面应与设计发展阶段相一致，而且在与不同的设计阶段相应的同时，价值管理可不止一次地实施。

初级阶段

价值管理可被用来作为决定项目是否真正需要的一种方法。在这种情况下，需要使用最高层面的功能分析。例如地方政府决定兴建一所新的发电站。功能分析的结果显示这是需要的，因为现有发电站不能满足用电需求。因此发电站的功能是"用电满足需求"。这种需求可以通过提高电力供应或者减少用电需求来实现。减少用电需求的一个方法就是鼓励人们使用节能灯泡。因此与其新建一所发电站，更好的方法也许是免费提供节能灯泡。

纲要阶段

一旦决定继续进行一个项目，那么可以用价值管理方法来建立纲要。此时适合使用空间层面的功能分析。通常在此阶段，仍有可能决定该项目是否需要。在一些价值管理作者和从业者看来，在此阶段应用价值管理方法建立纲要是最为有利的。例如，在老年疗养院里，病房的功能被定义为"允许停留"，"为护理提供便利"及"提供食品"。于是提供食品这一功能衍生出多种选择可能，如自助餐、轮椅餐、餐厅餐、从外面卖餐户或者在院内备餐等，所有这些选择都是可行的。

概述方案（35%的设计）

在此阶段设计得到了进一步的发展并且价值管理团队可以以分项设计和说明为基础提供多种选择方案。这时将会用到分项功能定义。在这个时候价值管理仍然可以提出对项目本质或者其空间布局的改变方案，但在此阶段这种改变方案会导致重新设计。

在施工期间

由承包商提出价值管理方案的做法已经有相当一段时间，尤其是在美国，在那里这些被称作价值工程变更方案。在这种情况下，承包商经常被给予经济奖励去制定方案，并且这种奖励通常以所实现的节约成本的百分比的形式出现。价值工程变更方案存在许多问题，最起码是设计责任的问题。另外，许多采用传统项目采购方法的承包商感到提供变更方案并没有给他们带来真正的经济利益，因为所得的经济奖励也许会与项目规模的减少相抵消。

如何评估可选方案

评估价值管理方案的方法有许多种，但最常用的是权重矩阵法，如表5.3的例子所示。

表 5.3 权重矩阵法

方 法	最初成本	维修	美观	能量效果	施工时间	合 计
权重	6	3	10	3	3	
白炽灯	18/3	3/1	20/2	3/1	6/2	50
水银灯	18/3	9/3	20/2	6/2	6/2	59
日光灯	24/4	12/4	10/1	9/3	9/3	64

第五章 价值管理(Value Management)

这个权重矩阵展示了一系列可能的灯具,以及对他们进行评估的一些标准。每一项标准都被加权以反映其重要性,就上面的例子来说,美观被认为是最重要的标准,成本是其次重要的,以此类推。然后根据下面的级别,对每种灯具的每项标准都给出一个等级值。

杰 出	5
非常好	4
好	3
一般	2
不好	1

在上例中,当根据成本标准进行评估时,日光灯具被评为非常好的等级(4),但是从美观角度而言它们被划归在不好的级别中(1)。然后,将所定级别与权重相乘,得到最初成本一项总分为24分和审美一项的总分为10分。然后将这些分数进行累加,此例中为64分,再与其他种类的灯具的分数进行比较。得分最高的方案为最佳方案。本例为日光灯具。

与此相似的另一种评估方法是SMART系统。它与上文介绍的系统略有不同,如表5.4所示,所有的分配权重必须是分数。另外,每种可选择方案的分数不是以预先决定的范围为基础,而是根据该可选方案满足标准要求的程度而定,并采取百分制。因此,在日光灯具的例子中,就其成本满足标准的程度而言,应为100中的80,而就其外观满足标准的程度而言仅为100中的10。之后,将这些分数与权重相乘,所得数值进行合计,来得出其总分排名。最高分数的可选方案将被认为是最佳选择。

表 5.4 SMART 方法

方 法	最初成本	维修	美观	能量效果	施工时间	合 计
权重	0.24	0.12	0.4	0.12	0.12	1.00
日光灯	80/19	80/10	10/40	60/7	60/7	83

其他的评估技术较少地依赖数量化,而更多的是依赖于主观评价。有些仅仅依赖于投票系统。量化评估发生的程度实际上依赖于项目发展的阶段。在纲要之前的阶段,可能很难对价值管理团队提议的可选方案进行量化分析,因而选择也许仅取决于业主的倾向。不过,如果可选方案是组织有序的,那么也许可以用权重矩阵法来分析。

价值管理作为一个系统

希望本章已经说明了价值管理技术的中心是功能分析。除此之外就是有效开展功能分析的方法。然而因为价值管理是个较新的理论,而且因为对于不同的人价值概念也不一样,所以目前还没有一个明确的系统,而仅仅是对可选方案的组成内容的选择。

表5.5对这些组成内容进行了概括。一套价值管理系统包括了为每个组成内容提供一个选项,以及这些成份如何组合在一起从而构成价值管理系统。公司应该选择最适合他们的价值管理系统及其具体内容。表5.5的总结并非完全详尽。价值管理仍处于发展过程中的早期阶段,而且随着对于系统内容及其用途和他们相互间发展的理解,该选择内容的清单还将扩展。

表 5.5 价值管理组成内容

组成内容	选择项
功能定义	基于项目功能 基于空间功能 基于分项功能
功能评估	实现功能的最低成本
功能分析系统技术图表	使用 不使用
按功能进行成本分配	是 否
计算值的系数	是 否
可选方案的产生	暴雨式思考 其他创意性技术
研究组织	工作计划
团队的方法	外部团队 设计团队 二者混合
促进价值管理员	独立的(外部的) 公司内部的
价值管理研究的形式	40h 研讨会 两天研究 其他适于该项目的方法
地点	工作环境之外 工作环境之内
研究的开始时间	项目初期阶段 纲要阶段 草图设计阶段 施工阶段 以上各阶段的综合 连续过程
对可选方案的评估	权重矩阵法 例如, SMART 法 其他数学方法 投票法 主观评估

下面列举了一些典型的例子来说明这些组成内容是如何组合在一起从而形成价值管理系统的。

美国价值工程系统

美国系统(表 5.6)是建立在 40h 研讨会基础上的,而且该研讨会由外部小组在设计完成了 35%的阶段召开。研讨会围绕工作计划展开。美国的价值管理的实践基本上是以分项功能这一层面为基础,并围绕分项功能产生多种可选方案。

第五章 价值管理(Value Management)

表 5.6 美国的价值管理系统

组成内容	选择项
功能分析	基于分项功能
功能评估	实现功能的最低成本
功能分析系统技术图表	采用
按功能进行成本分配	是
计算值得系数	是
可选方案的产生	其他创造性技术
研究组织	工作计划
团队方法	外部团队
价值管理促进员	独立的
价值管理研究的形式	40h 研讨会
地点	工作环境之外
研究开始的时间	草图设计阶段
对可选方案的评估	权重矩阵法

一个美国的价值管理案例分析

研 究

该项目是国防部的一栋培训楼,预算成本为 240 万美元。价值管理促进员是一位来自价值工程咨询公司的有资历认证的价值专家,并有着土木工程专业的背景。价值管理团队是外部团队,由一个建筑师、一个机械工程师、一个结构工程师和一个电气工程师组成。他们是由该促进员从其他咨询公司中挑选出来的。项目处于设计完成了 35% 的阶段。这是一栋一层的钢筋混凝土楼宇,采用打桩基础,混凝土地面,金属屋顶。这栋楼包括一个单轨铁道和提升机、音响设备、高压机、排气系统、防火系统、空调和其他设备。价值工程团对提交的改造的方案的造价为 535 980 美元中总共节省了 154 000 美元。该价值管理研究成本大概为 21 162 美元,因此得到 7.3 的投资回报。该研究采用 40h 研讨会的形式,并对项目做出了以下调整方案:

价值管理方案

- 减少了传音的中央机械主机的数量
- 取消弹性地板和密封混凝土
- 取消吊顶天花板和油漆
- 取消缓冲空间
- 保留外部分隔
- 减小塔吊承重梁跨度
- 将大跨度屋顶托梁改为 K 型系列
- 采用钢架双墙结构

- 采用角钢托梁
- 修改外墙的控制托梁
- 取消回气管
- 增加关闭阀门
- 改进送气、排气口细部
- 应用要求和分流系数,拆除 MDP 板
- 减弱天篷照明
- 重新布置屋顶通讯系统
- 重新布置停车场照明
- 减小人行道宽度
- 修改泻洪排水系统
- 改变沥青铺地的种类
- 改变混凝土路堤的种类

英国价值管理系统

到目前为止在英国还没有确定的系统。表 5.7 所列的方法是约翰·凯莉(John Kelly)所使用的,并且有关的记载表明此方法得到了认可。这种价值管理系统比美国的价值管理系统要早,并且通常由项目设计小组实施,而由外部价值管理促进员领导。功能分析被用来理解项目的目的,和产生可以改进项目的多种可选方案。

表 5.7 英国的价值管理系统

组成内容	选择项
功能定义	以项目功能为基础 和/或者以空间功能为基础
功能评估	实现功能的最低成本
功能分析系统技术图表	采用
按功能进行成本分配	是
计算值得系数	否
多种可选方案的产生	创造性技术
研究组织	工作计划
团队方法	设计团队
价值管理促进员	独立的
价值管理研究的形式	两天的研究
地点	工作环境之外
研究开始的时间	纲要草图设计阶段
对可选方案的评估	权重矩阵,例如 SMART 法

一个英国的价值管理案例分析

该项目是一幢公共楼宇的重新装修,价值管理研究的目的是为项目建立控制纲要文件。研究及所得纲要包括了下文中列举的各个不同部分。

对策略性问题的调查

研究中涉及到的策略性问题包括对从工作人员到来访者等楼宇使用者进行的调查。这包括对于来访者类型,他们的需求以及他们的等待行为的分析。其他策略性问题还涵盖了项目发展的政治背景、社区服务、现存楼宇、未来技术变更潜能、资金、安全以及保卫等。价值管理团队提供的策略性建议包括:

- 将楼宇中的工作人员与那些在另一幢较小楼宇中的人员安排在一起。
- 协调维修及资本预算以实现更好的价值。

对于时间,成本和质量的重要性的分析

在这个部分的研究过程中,价值管理促进员向团队成员提出这样的问题:从时间、成本和质量的角度考虑项目的主要优先项是哪一项。工作组总结道:时间最不重要,而成本和质量同等重要。

功能分析

功能分析系统技术图表被用于帮助功能分析。功能分析系统技术图表以功能定义中的最高层面为起点,而且概括了项目的目的是反映一种公司方法。此目的在功能分析系统技术图表的基础上一直贯穿下去,直至诸如卫生间和母婴室等楼宇空间功能的层面。与美国的 FAST 图表不同的是它没有试着显示空间和分项之间的内在联系。该图表的用处在于它显示了建筑空间之间以及其与更高层面功能之间是如何联系的。

用户及用户流动图表

价值管理团队制订了该楼宇的使用流动图,包括保安员、打字员、接待处、会谈室、普通职员、财会职员及通信员等。该流动图显示了各部门将如何使用他们的办公室,其目的在于提高使用效率。

空间定义

空间定义检验了项目提供的主要的空间,并特别关注空间质量和空间环境。其目的在于检验质量和环境是否与楼宇的本来使用意图相一致。

对于空间位置及比邻空间的研究

通过使用矩阵法,并基于上文中提到那些对该项目已做的研究工作,价值管理团队建立了比邻矩阵,这个比邻矩阵显示了楼宇中所有相关空间的位置。为了达到最大空间效果的目的,加之管理团队给予每个空间 0~5 的分数,这里 0 代表无相邻要求,5 代表有明确的相邻空间的需要。

一套签合同前的计划书以说明主要行动点

这一部分很大程度上是自我解说,并与其他签约前计划有相似之处。其中指出的主要行动点是关于设计者的方案提纲、成本计划、数量清单和投标时间。

对于需要立即采取行动的项目的具体内容的研究

基于上文所列举的所有实践方法,价值管理团队列出了一张需要马上实施的清单,共包括 21 项,其中包括:
- 检查对于地下室防火疏散的要求
- 解决地下排水(问题)
- 确定楼面荷载
- 实施安全检测
- 重新检验等候室和会议室的大小
- 重新检验接待台和接见室的数量

研讨会的议程

按照以下议程安排,研讨会持续了两天:
第一天
- 价值管理介绍
- 促进员介绍信息
- 策略问题研究
- 午餐时间
- 时间、成本和质量研究
- 时间安排表
- 功能分析

第二天
- 回顾前一天工作
- 楼宇用户流动图表
- 午餐时间
- 功能性空间定义
- 质量和环境研究
- 比邻空间
- 回顾及闭幕

价值管理团队

该团队包括以下成员:
- 4 位来自业主的成员,包括 3 位工料测量员
- 7 位职员或楼宇使用者
- 2 位来自建筑师事务所的建筑师

第五章　价值管理（Value Management）

- 1位机电工程师
- 2位价值管理促进员

这个英国案例的分析介绍了价值管理在英国的发展状况。虽然功能分析仍是价值管理研究的核心部分，但是研讨会还包括了更广泛的研究内容，即对于项目目的的检验和使用者的需求的研究。另外，与美国的价值工程研究不同的是其对于成本的强调不多，而是将重点放在提高楼宇的使用性上。不幸的是，上文中列举的实例仍很少见。不过正像本章前面部分提到的一样，英国的价值管理正处于其发展的十字交叉路口上，也许这卷研究很快就会普及开来。总而言之，这个价值管理研究会议开了两天，而且价值管理团队的成员除促进员以外，都是那些已经参与该项目内容的人员。因此成本相对较低，而且这项研究对整个项目所带来的利益还是不错的。

其他英国案例分析可以在格林（Green）和伯坡（Popper）的著作中查到。

日本的价值管理系统

与美国和英国的系统不同，日本价值管理不是易而为之，而是在建设项目中实施的一种持续过程。日本更多地将价值管理看作是一套哲学而不是一个系统，而且在建设周期中的全部阶段，包括规划、维修和环境保护中进行运作（表5.8）。

表5.8　日本的价值管理系统

组成内容	选择项
功能定义	基于项目功能 基于空间功能
功能评估	实现功能的最低成本
功能分析系统技术图表	采用
按功能进行成本分配	否
计算的值系数	否
多种可选方案的产生	其他创造性技术
研究组织	工作计划
团队方法	设计团队
价值管理促进员	内部的
价值管理研究的形式	其他可实施方法
地点	工作环境之内
研究时间	持续过程
对可选方案的评估	主观评估

一个日本的价值管理案例分析

在日本，政府的和私人的项目采购系统之间非常不同，而且这种不同对价值管理的实行也产生了影响。私营企业经常采用设计连施工合二为一的采购方式，而且在这种情况下，价值管理的研究也由建设公司自己进行。政府/公众的企业还在使用传统的设计和施工相分开的方式，在这种情况下，价值管理和研究相应地分为两个阶段进行。首先先由公司内部的设计师和顾问在设计进展的过程中进行，然后由施工公司对项目进行价值管理研究。价值管理已经成为项目采购过程中的一个组成部分。下文列出价值管理的常用过程。这些内容摘自可波(Kobe)城市房屋公司的项目，一个包括地下室的 17 层高的楼，该楼设有图形图像的多功能厅，楼面总面积 14 261m²，总造价为 410 亿日元。对项目的价值管理研究程序如下：
- 第一步：决定哪些承包商可以参加投标
- 第二步：给这些承包商提供设计师的设计图纸，概算清单和价值管理的程序
- 第三步：这些承包商在现有设计图纸的基础上编写价值管理方案
- 第四步：业主(委托方)对这些价值管理方案进行评审并将之归类为好的或坏的两类
- 第五步：这些承包商从中选择几个被评为好的价值管理方案，而且编写最终方案并与最终报价一起提交业主
- 第六步：业主选择最低报价的承包商中标。这个最低报价必须不多于选择参加投标的承包商时的原始价

上述项目中承包商提交的价值管理方案一般包括：
- 将第 5 层以上的框架由现场浇灌改为预制混凝土
- 将第 5 层以下的模板由木模板改为金属模板以增强重复使用次数
- 使用回收的木材而不用全新的
- 改变屋顶管道的位置
- 改变消防系统的管道布置

上述例子并不是什么标准的价值管理系统，因为在日本有很多不同的实行价值管理的方法。在一些情况下，提出价值管理方案的中标的承包商会得到所节省造价的一个百分比。

为什么这些系统互不相同？

价值管理系统是彼此独立发展的，因为各自所在的建筑业及商业文化环境是不同的。对于这些文化背景的研究使我们清楚了解这些系统的特点。本书中同样提到，当从其他国家引进建设管理技术时，必须考虑到商业文化背景的影响。对于这些文化差别的调查研究为那些希望发展一套适合他们自己的商业文化的价值管理系统的企业提供了解详情。商业文化背景对于价值管理的影响很大，下文阐述在美国和日本，它是如何影响价值管理的发展的。

管理类型的差异

在信息汇集过程中，美国管理者很重视尽早采取行动，并满意于快速产生的结果。另一方面日本管理者重视概念，并以长远眼光看待问题；他们并不拘泥于固有的解决问题的方法，而是从多种选择性中寻求最满意答案。这说明在美国，人们已经满意于一套会导致特定的结果的固定的价值管理系统。而在日本，更具扩展性和进化的过程可能更为适合。这在两套价值管理系统中得到了真实的体现；美国模式提供了一套相对简短而固定且会产生切实的结果的过程，而日本模式则为持续性的和不那么固定的。

从信息评估角度而言，美国管理者重视逻辑性且最满意于具体的解决方式。另一方面

日本管理者重视此阶段上的人际互动，很大程度上基于过去的经验，满意度并非来自于解决方法的产生，而更侧重于对问题的整体评估。对于由价值管理产生的概念的评估而言，美国系统更倾向于精确的数学统计方式，而在日本直观感觉性的系统可能更为适合。

管理体系的不同

管理方式导致了管理体系的不同。日本与西方的管理体系是不同的而且这与价值管理的运作有直接关系。表 5.9 总结了这些管理体系的主要差别。

表 5.9 不同管理体系之间的差别

日本管理依靠于	西方管理依靠于
团 队	个 人
长期目标定位	短期结果
职责的不明确	清楚定义的职责与任务
人文主义管理	逻辑与理性管理
利用人	利用功能
和 谐	公 平
接受不平等的关系	平等的关系
不断的进步	结 果
对年龄和辈份的尊重	在业务关系中年龄和辈份不太重要
一致意见的基础上做决策	由资深者做决策
由保留脸面控制	由上级控制
以任务的方式进行控制	意识观念上的控制

通过检验不同国家的价值管理系统，可以看到不同的管理系统是如何影响这些价值管理系统的。对于英国和美国的价值管理系统而言，问题最大的领域毫无疑问的是人际关系，这很大程度上是由于管理系统依赖于个人的原因。不像日本，团队行为不是固有的，因此颇具问题。对于短期效应的要求使得美国系统具有很强的结果导向性，绝大多数的价值管理研究都计算出一个投资回报率来显示对该研究的成本投入所达到的回报程度。大量计算利润回收的研究显示了与价值管理研究成本有关的可实现盈余。这一目的使得价值管理工作组试图去发现可能实现的更大的盈余，而忽视了功能分析。对于逻辑性和理由的强调还导致了美国方法比日本方法更具确定性，与之相比日本方法虽然清楚完整，但却缺乏确定性。另外，任何基于长期规划基础上的商业文化都有可能更加接受对立于成本的价值概念。这相对于由任务控制的系统而言，被意识形态所控制的系统也如此。

价值管理与造价工程间的关系

本章已经说明，价值管理从本质而言与传统造价工程师之间没有密切关系。这并不是说造价工程师没有能力将其传统业务服务范围扩展到价值管理领域，他们正在这么做而且这方面的实践活动已经提供了价值管理服务内容。然而价值管理的发展并不属于造价工程师的领域，因为造价工程师比任何其它专业人员都更适合于价值管理促进员这一角色，这是因为皇家造价师协会(Royal Insstitute of Chartered Serveyers-RICS)对于价值管理的发展做出了积极的促进作用。

结　论

　　用评估功能这一概念来作为评定价值的一个方法是存在问题的，理由有二：第一，价值是主观性的；第二，价值会随着时间而改变。

　　一件珠宝首饰，例如结婚钻戒，只具备非常有限的功能。它也许是为了表明佩带者已婚，或者仅仅为了达到修饰作用。但是对于拥有者而言，尤其当他们拥有这件珠宝很多年的时候，其价值也许远比用功能表示的价值高，因为它已具备了情感价值。我们大多数人都会承认情感价值的存在，并且意识到此价值虽然真实存在，却不能定义或量化。

　　除了其主观性以外，事物的价值还会随着时间而改变。正如本章开篇时提到的，当暴风雪切断了电力供应时，用电池操作的收音机的价值可能得到很大程度的提高。再举一个不太具戏剧性的例子，商品的价值，例如儿童足球的价值会因新型产品的上市而被降低。

　　因此我们还没有真正的衡量价值的尺度，而只是一个在某一时间，对某一个体或某一群个体而言的价值的评定。然而我们有的，并且总可以达到的，是对我们所需要的或想要一件东西所能实现的功能的定义和评估。在这里，需要和想要之间的区别是很重要的。在价值管理中，功能不仅是指项目需要做什么，而且也是我们想要它做什么。如果只有考虑需要，那么所有楼宇可以是一个有平屋顶和地毡地面的四方体。然而很明显，许多业主都希望他们的建筑物能提高公司的声望和提高舒适的环境，这些也可以同样被定义为功能。

　　在建筑业中，功能分析的一大难题是，很难决定对谁的功能和价值进行评定。对于支援建设一家新医院的政府来说，新医院的价值也许是增进该地区的政治支持。对卫生部门来说，其价值是对病人和其职员提供更好的设施。对工作职员来说，其价值是改善了的工作环境，而对当地居民来说，是该地区拥有更好的设施因而减少了延误和路途的时间。这其中首要的是对社会的价值，即提供更好的基础设施，改善所有人的生活质量。由此产生的问题是，在所有这些主观的和变化的价值中，应该评定哪一个。出于价值管理的目的，答案是对所有这些都要评定。

　　但是是否有可能对于医院项目中涉及的所有的需要和想要进行评估，从政客到病患者？从价值管理角度而言，答案是我们将那些对该楼宇感兴趣的人们的需要和想要转化为功能；也就是使用者需要什么且希望楼宇如何满足其要求。这并不是件容易的事，并且需要采用一定的技术将该楼宇的真正功能从相互抵触的众多意向中整理出来，然后提供能够满足那些功能要求的最为经济的设计方案。

　　虽然功能分析是价值管理的重心，但是要使功能分析能有效实施，其它组成部分同样是需要的。在这些其它组成部分中，包含多种可选择方案，并且对于这些构成价值管理系统可选方案的选择要依靠许多因素，例如项目属性、时间规模、设计团队和行业的商业文化背景。因此没有哪一套价值管理系统是绝对正确的，因为最为有效的系统将随着不同项目的情况的变化而变化。所以灵活性是价值管理的关键。

　　对成功的价值管理来说，理解价值管理的组成部分及其最佳的应用时间是至关重要的。然而价值管理是一个相当新的技术，对于在什么时间使用什么方法的资料并不是总能得到的，而且也许很长时间内都不可能得到的。在解决这些问题之前，价值管理还需要更多的时间和研究。

第五章 价值管理(Value Management)

对价值管理来说,最大的一个错误就是将它与减少成本相混淆。我们希望此章已经阐明,实际上价值管理与成本只有很少关系,它是一个设计过程。事实是,成本的减少通常是价值管理的一个结果,而不是它的目标。

第六章

可建设性（Constructability）

导　言

在任何一本标准字典中都找不到"可建设性"（Constructability）和"可建筑性"（Buildability）这两个术语。它们是建筑业的专用术语，并且仅限于在此行业范围中使用。公平的说，虽然这两个名词所代表的原则越来越被更多的国家所接受，但是，"可建设性"和"可建筑性"这两个词在许多从业者词汇中的使用仍不常见。

在本章中这两个词是同义的，使用上可以互换。鉴于一致性上的考虑，我们在这里选择使用"可建设性"一词，除非引用他人文献时而在其原文使用了另一术语。我们已经回避了在两词中存在的微妙差异，忽略一些民间说法，例如可建设性是英国用法，而可建筑性是美国用法，或者可建设性包含的系统范围比可建筑性更广等。

与本书中涉及的所有其他概念相比，可建设性的行业特性使其成为独一无二的概念。诸如全面质量管理和重组等概念涉及了一系列不同行业领域，而可建设性与众不同，它是建筑业在过去30年中为其自身行业创造并发展起来的唯一的管理概念。

起　源

与其他行业相比，设计与施工过程的分离是建筑业中独一无二的现象。这种功能上各自为政的划分长期以来已在一些报告中被强调，例如西蒙（Simon）的报告，埃默森（Emmerson）的报告和班威尔（Banwell）的报告。针对这些已被认知的缺陷，建筑业研究及信息协会（CIRIA-Construction Industry Research and Information Association）于1983年将注意力集中在可建筑性这一概念上。建筑业研究及信息协会认为确实存在可建筑性的问题，"也许因为多数设计师与施工实践过程脱节的原因。在施工人员看来，这些问题不是某个人的错，而是因为设计与施工功能的分离，而且这种情况一直是英国上个世纪建筑业的特点。建筑业研究及信息协会将可建筑性定义为"在满足楼宇竣工后所需的一切要求的前提下，楼宇的设计使得施工容易的程度。"建筑业研究及信息协会的这个定义只是集中在设计和施工的关系上，并且暗示了仅仅那些设计小组所能影响或者控制的因素正是那些对于楼宇施工难易程度有显著影响的因素。

与此同时，美国成立了建筑业协会（CII—Construction Industry Institute），该协会的主要目的是提高经济效益，实行全面质量管理，提高美国建筑业在全球的竞争力。可建设性一直是美国建筑业协会研究发展工作的重要组成部分。美国建筑业协会对于可建设性的定义范围较之英国的建筑业研究及信息协会的更为广泛，并将之定义为："一个在项目规划、设计、采购及实际施工等一系列过程中实现最优化的施工知识与实践经验二者的最佳结合，并使各类项目与自然环境条件限制达成平衡，以实现项目目标的体系。"

20世纪90年代初，澳大利亚建筑业协会(CIIA—Construction Industry Institution, Australia)将美国建筑业协会的可建设性过程加以发展，使之更适用于澳大利亚情况。因而，澳大利亚建筑业协会将美国建筑业协会的可建造性的定义修改为："一个实现施工知识在整个建设过程中的整合，并使各类项目与自然环境条件限制达成平衡，以使项目目标与楼宇的表现得到最大体现的体系。"

可建设性的目标

可建设性的目标由其定义的范围来决定。1983年英国建筑业研究及信息协会将可建设性概念的范围限制在设计与施工的相互关系上，参见插图6.1。

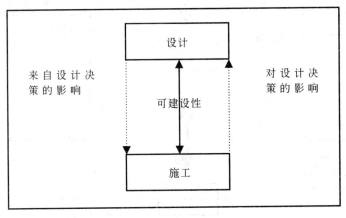

图6.1 英国建筑业研究及信息协会定义的可建设性的范围

在系统术语中对概念模式范围的描述被认为是系统界限。英国建筑业研究及信息协会的模式的系统界限非常窄，仅仅将可建设性看作是与设计有关的行为。格力菲(Griffith)认为这种方法实际上严格限制了这一概念，并导致可建设性丧失了在英国实施的动力。一些研究人员曾讨论过取得可建设性模式适当界限的困难。另一方面，如果界限过宽，在实施过于简单的方法的同时就会伴随着一种固有危险，即将可建设性等同于其实施前景非常渺茫的一套本体陈述。反之，如果实施一套集中局限的方法，那么又不能完全实现概念的潜能。

一个可行的可建设性概念需要认识到在项目环境中有很多因素，它们会影响设计过程，施工过程，和设计与施工间的联系和楼宇的保养维修工作。这可以通过图6.2加以说明。

图6.2说明在一个项目环境中有许多会影响设计过程，施工过程，楼宇的质量和效果的因素。只有认识了这些因素间复杂的相互作用，才能发挥可建设性的潜能。重要的是不但要认识到可建造性并不等同于容易施工而且要认识到可建设性也关于楼宇建成后的合适性。这可以从可建筑性的一个定义中看出来，即"可建筑性是贯穿整个楼宇采购过程中所作决策的范围，它反映了影响项目目标的因素，最终使得施工容易，并提高项目竣工时的质量。"(此定义已被新南威尔士州政府的施工政策指导委员会采用。)另外，对可建筑性的关注并未因工程竣工而停止。楼宇维修方面的可建筑性，例如在安装、拆除和材料更新，楼墙面，管线及设备在整个楼宇使用过程以及在施工初始阶段的可建筑性，都是同等重要的。

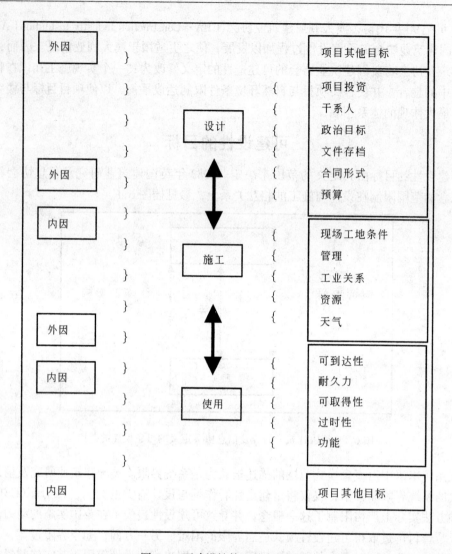

图 6.2 可建设性管理的实施框架

设计阶段之前所作的决定将对设计决策过程增加约束。与此同时，那些不是由设计师所作的关于设计和施工之间的关系的决定也将对施工产生重要影响，比如文档，承包商选择等等。与此类似，维修管理员通常被看作是设计和施工决策过程后面的，将会受到与建设性决策的显著影响。参见图 6.2，可以看到项目的每一个阶段，从设计到使用，都受到诸如外因、内因和项目具体目标等因素影响。（外因是指来源于外部的因素或是由于外部压力的影响。内因是指源于内部或由内部发展出的因素。）一个考虑了外因，内因以及项目目标的方法必然会面临一些复杂的问题。如果采购过程的复杂特性没有被意识到，那么将存在着可建设性可能变成无实际用途的普通策略的危险。

实施可建设性

可建设性原理

上一节中我们已强调了如果要实现可建设性的目标，就需要以平衡的视角观察事物，并建立适当的系统边界。这个问题已被澳大利亚建筑业协会强调并曾与美国建筑业协会协力制定了最佳实践，如何去做可建设性手册。（对澳大利亚建筑业协会关于可建设性原理的详细解释和一般原理，读者可以参阅格力菲（Griffith）和西菲尔（Sidwell）的报告）。美国建筑业协会和澳大利亚建筑业协会的可建设性文件不仅为从业人员提供了详细的清单，而且与可建设性的概念模式（图6.2）保持一致。以下是澳大利亚建筑业协会的可建设性原理文件的目录：

- 企业如何建立一套可建设性方案的执行性建议；
- 项目周期中不同阶段可建设性原理的适用性的流程图；
- 可建设性原理的概要；
- 十二项可建设性原则；
- 用于记录可建设性中节省造价的范例的数据库。

澳大利亚建筑业协会提倡一套有结构的方法，这套方法将楼宇采购过程划分为如下五个阶段：

- 可行性
- 初步设计
- 设计
- 施工
- 施工之后

十二项原理内容如下：

1. 一体化——可建设性必须作为项目计划的必要组成部分；
2. 施工知识——项目计划必须主动包含施工知识与经验；
3. 团队技能——项目团队的经验、技能，以及人员的构成必须与项目的属性相适合；
4. 共同目标——当项目团队与客户项目目标达成共识时，可建设性将会得到提高；
5. 可用资源——设计方案的技术要求必须与已有技能及可利用资源相匹配；
6. 外部因素——外部因素会影响项目的造价和/或者项目的进度；
7. 方案——项目的整体方案必须是可实现的，易施工的，且有项目团队的许诺；
8. 施工方法——项目设计必须考虑施工方法；
9. 可达到性——如果在项目的设计及施工过程中考虑到施工的可达到性，项目的可建设性将得到提高；
10. 具体方案书——如果在项目具体方案书中考虑到施工效率问题，项目的可建设性将被加强；
11. 施工技术创新——施工过程中创新技术的应用将加强项目的可建设性；
12. 反馈——如果项目团队在完成施工之后对该项目进行分析，将会提高未来同类项目的可建设性。

表6.1说明了澳大利亚建筑业协会12项原理在项目采购过程的5个阶段中的分配情况。这些原则依据其在项目采购阶段中的地位的重要性被划分在三个级别中。例如外因在可行性阶段中非常重要[6],(6)在初步设计阶段比较重要,但是在具体设计阶段不很重要,在施工及施工之后阶段则毫无作用。这些级别是对美国建筑业协会提供的原则方法上的使用指导,而且会改变的。它们只是帮助使用者确定哪些原则可能与项目生命周期的特定阶段有关或者对其至关重要。

表6.1　12项原理在项目采购过程中的分布情况

可行性	初步设计	具体设计	施　工	施工之后
[1]	[1]	1	1	1
2	[2]	2	2	4
[3]	[3]	[3]	(7)	[12]
[4]	[4]	(5)	(9)	
[5]	(5)	6	[11]	
[6]	(6)	(7)		
7	[7]	[8]		
8	[8]	(9)		
	[9]	[10]		

注:[　]极为重要,(　)比较重要,没有括号代表不那么重要。

在项目进展过程中,不同的人相对于这12项原理有不同的作用和责任,并且每个人在整个项目周期的不同阶段会有不同的职责。因此每项决定都需要协调以便可建设性在项目中得以顺利实施。否则,个人为在其影响范围内达到目标可能采取不同的策略,这样将会削弱项目整体的可建设性的实施。

图6.3展示了可建设性实施计划的框架,它定义并协调整个项目周期的各个项目参与者的决策角色和职责。这样可以使得每个项目都有可建设性计划,这样就不仅允许个人定义其自身作用及职责,而且允许他们看到其他参与者在项目周期的每个阶段正在做的和应该做的,从而更好地促进了工作的一体性和协调性。

此方法的精华是通过各参与者运用他们在施工方面的知识来获得最大的机会,以协调的方式发展最佳方案来实现项目目标,而且通过采用例如价值管理等回顾过程的方法,从而加强可建设性。

实践中的可建设性

为了使可建设性得以成功应用,业主或者业主代表首先应制定一个方案来清晰详尽地阐明项目的主要目标,并将可建设性作为评估项目表现的内容之一。可建设性的目标也应该在项目团队中所有成员的不同作用和职责中明确定义。这可以通过实施那些表达了业主和使用者需求的关于时间,成本和质量标准要求,以及使用图6.3中的框架得以实现。图6.3中的框架协调了项目团队在整个项目周期中对可建设性原理的考虑。

虽然,设置一种可在项目团队成员之间协调可建设性原理的机制是实施可建设性的一个非常重要的方面,但是,对于由项目团队各成员确定在项目采购过程中参与的时间,也同样重要(图6.4)。

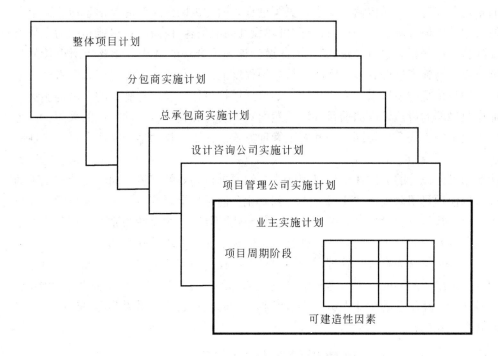

图 6.3 可建设性实施计划框架

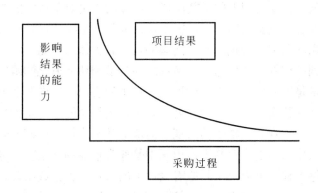

图 6.4 成本影响/帕累托(Pareto)曲线

帕累托(Pareto)原理对时机的重要性进行了阐述,该原理强调,在项目周期的早期阶段做出的决策较之在后期阶段做出的决策将会对项目的最后成果产生更大的影响。"在项目早期阶段,诸如初步规划和设计阶段中,决策对于项目成本的影响很大,但是随着项目向施工阶段的发展,其影响会快速减小。因此,从最初阶段起对于项目不同设计和施工方面做出的决策则显得至关重要。尽早让承包商介入项目被看作是获得良好的项目可建设性的理想途径。"

在此是考虑设计连施工的采购形式和可建性间的关系的一个好时机。毫无疑问,设计连施工的采购形式有助于承包商尽早进入决策过程。因为英国建筑研究及信息协会对于可建设性的最初的兴趣来源于一个观点，即"可建设性问题是客观存在,这可能是因为多数设计者与实际施工过程相对脱节导致的"。因此设计连施工合一的方法可能要经历很长的阶段才能解决设计与施工过程脱节的问题。从某种程度上说这是事实，并且多数学者都赞成这一观点，即与传统的项目采购方式相比，如选择性招投标，设计连施工的方法确实提供了一个促进"良好"的可建设性的项目环境。然而对于设计和施工合一的方法的采用，或者由此而衍生出来的方式，如合同管理和施工管理方式，并不会自动地带来适合可建设性的更好方法。

成功实施可建设性的关键在于项目团队成员间要进行高效的交流，同时设计连施工的采购方式可以使项目团队成员的交流方式得到简化，但是如果要获得成功，项目团队成员自身必须认可和执行可建设性。有一种观点认为导致"不良"可建设性的原因一部分是由于对现代建筑设计者的教育重点的转变，并且对于建筑学课程中建筑施工部分没有给予充分的重视。另一种观点认为在施工管理课程中对管理课程的内容的重视不够，或者是对于设计鉴别的内容未给予足够的重视。这种争论是无休止的，并且通常会受专业背景所影响。问题的关键在于在建筑业的专业人员获得多专业合一的教育之前，阻碍获得"良好"可建设性的障碍总会存在的。

可建设性和楼宇产品

可建设性的首要目标就是利用一切可利用资源建造出最好的产品，也就是楼宇。虽然如此，我们还惊奇地发现，很少人关注可建设性与在使用的楼宇间关系。这是因为我们忽视了一个事实，也就是在项目周期的最初阶段，所做的决定对于楼宇的维修与更换以及组装和拆卸的简易程度有着重要的影响。研究工作探究了楼宇采购初始阶段所做的决定和这些决定对于投入使用的楼宇的表现的影响间的关系。其大体总结如下：

楼宇是经常受到结构方面在生命周期结束之前技术和功能退化的威胁的固定资产，而这些退化的发生通常是因为材料物理性能退化，技术的衰退，标准及功能变化等方面的原因。楼宇的局部维修和替换是楼宇生命周期中几乎每天都会发生的正常现象。因此不应在楼宇生命周期的某一阶段判断该楼宇的表现，而应对其整个楼宇周期的表现进行评价。楼宇的可能延长其有效生命的潜力在其生命周期中是非常重要的因素。一幢建筑物在其生命周期内的使用效能绝大程度上取决于其原有设计上的特点，施工或组装过程，以及正常运作，维修和最终拆除所产生的要求。项目初始阶段所做出的决策的质量很大程度上反映了该楼宇的表现水平。

楼宇的有效维修和更新的意思是容易对楼宇部件进行拆卸、维修或者替换。鉴于此，想办法让楼宇表现得更好应被看作是可建设性范围之内的。可建设性通常被看作是与楼宇的最初的建筑组装的容易程度有关的纯设计问题。这一想法限制并分散了项目早期的决策方向，并且使决策离开了对楼宇在其整个生命周期中表现程度的问题的考虑。

促进楼宇表现的关键在于有效的信息管理，特别是在项目早期阶段的决策对于项目结果有着显著的影响的阶段。决策的质量的提高可通过确定关键性的问题和及时的和相关的信息来实现。

对于以可建设性为中心的维修和更新管理，对项目政策制订人，设计者和其他干系人

第六章　可建设性(Constructability)

给予的有价值的决策支持，来自对主要决策环境的系统的确定及跟踪，并根据这些进行计划来实现项目预期目标。图6.5阐释了一个可以促进的以可建设性为中心的维修和翻新管理策略的项目决策支持系统框架。

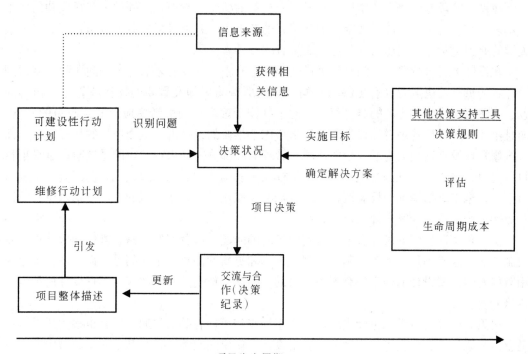

图6.5　以可建设性为中心的管理的项目决策支持框

该框架从概念上包括了以下功能：
- 行动计划(以可建设性为中心的行动计划)
- 完整的决策记录(项目综合描述)
- 信息获取、过滤和传递
- 获取相关决策工具和评估技术
- 项目参与者之间的交流与协调

以可建设性为中心的行动计划将确定那些对于可建设性将产生主要影响的问题并且提醒相关决策者那些需要他们注意的问题。这些行动的时机由贯穿项目周期中的每个阶段的整合的项目描述来控制。这个项目描述包括了已做出的所有有关该项目的决策。在可行性阶段，项目描述可以是对所有已认知的项目目标的陈述，而且这些目标逐渐发展成为设计纲要。在项目周期的不同阶段，项目描述可能会以诸如草图设计图，说明书，数量清单，以及施工图等形式呈现。实际上，项目过程中不同功能的各自为政的划分试图将由不同参与者综合起来的项目描述分割开来。这导致了交流与协调上的问题。计算机综合应用工具以及网络的应用将允许让所有项目参与者都清楚了解整体项目描述的发展。对于由行动计划产生的决策状况所做出的反应的质量将决定项目日后的表现，并且会被所能获得的相关信息，专家意见，及用于阐明、评价和挑选最佳解决方法的工具等加强。因为每个决策都被遵照执行，项目整体描述将被持续更新，而且知会所有项目参与者有关项目的信息，并且引发行动计划向下一步决策形势发展。

在设计阶段考虑楼宇的维护已不是一个崭新的概念。然而在此领域中获得成功的例子并不多。在项目早期阶段对楼宇维护和更新的考虑由于各种各样的原因被阻止，而其中最关键的原因包括：时间压力，无意识决策者和设计者。

在设计的初级阶段，时间是最关键的而且还有很多互相有矛盾冲突的问题都希望得到设计者的重视。决策者没有足够的时间来研究那些简单却又重要的操作或管道电器等问题，更顾不上维修和更新的难易程度。对于维修管理和更新的考虑应上升到政策的高度，而且由业主或项目管理者加以强调。那么考虑了维修的设计就会成为项目的目标之一，而且维修管理及更新则成为了一个整体项目管理的一部分。如果在项目的早期阶段定义项目目标之时，将楼宇的维修管理作为一个目标而被明确建立起来，那么就要调配资源来进行对于建筑未来维修情况的检验调查。这意味着要雇用不同技术领域的专家包括水电方面以及更换维修方面的专家来评判设计方案。

由于负责新设施的设计者未考虑到施工后的问题，以及那些维修、操作和保养这些新设施的人的经验不足，因此限制了对楼宇维修和更新的注意。不幸的是，设计者在楼宇使用阶段对他们所设计的楼宇的介入是有限的，因此，他们并不了解因为不良设计所导致的维修问题。

决策者就算已意识到可能出现的与楼宇维修和更新有关的问题，通常也很少能够有渠道获得可以辅助项目决策的正确信息。由设计者做出的决策通常是基于三个领域中获得信息：个人知识和经验，正式的参考资料以及当时的项目描述。在建筑业中，有一种倾向，即实践者会依赖于他们个人或同事的知识。这种做法虽然在多数情况下是有效的，也通常伴随着一些严格的限制。容易获得正式的相关的，有恰当索引和记录的参考资料的简易途径将为设计者做出恰当决策提供可靠的基础。

决策者需要获得必要的决策支持包括获得相关信息、专家意见和工具。在图 6.5 中提到的模式说明了如果关于运作/维修/更新阶段的问题得到了考虑，那么诸如生命周期成本，维修记录，后期使用评估和资产注册等决策支持手段都可获得。一套系统性的方法可以确保项目生命周期中不同阶段产生的要求及许多被割裂开来的项目决策过程的投入被统一综合在一起。侧重可建设性的整体管理策略为一个综合性框架提供了逻辑连贯性，该框架并未取代个体项目决策者的作用，但是却成为了针对全部项目参与者的共同的决策支持系统。这个计划框架还通过信息技术的实施，来解决那些发生在多数项目决策过程中都会出现的交流，合作及信息管理方面的问题。

使得楼宇维修和更新途径的相对容易是贯穿楼宇整个生命周期中的显著问题。侧重可建设性的的策略能够为提高维修和更新的容易提供逻辑性手段。这是通过对项目早期阶段提供相关决策支持的方法来实现，因为项目早期的决策对于整个楼宇的使用表现有着显著的影响。图 6.5 所提出的信息管理框架为所有项目决策者提供了一套独立的决策支持方法，并且考虑了项目整个生命周期中的动态要求。信息技术的使用使得复杂的协调与交流的要求得到满足，从而克服了各自为政的问题。

[注：以上描述是基于澳大利亚纽卡索大学楼宇表现研究小组正在进行的研究的基础之上，该研究

着重于识别侧重可建设性的管理中的重要因素。]

好的与坏的可建设性

可建设性已经被反复强调是一个因项目而异的最为活跃的工程要素。虽然将好的和坏的可建设性的实例记录下来，并以此建立和扩大可建设性的知识库是很重要的，但认识到将一个项目中好的可建设性范例应用到另一个项目中的局限性也同样重要。在过去，尤其在英国，许多有记载的案例研究似乎都着眼于失败的地方，也就是着眼于坏的可建设性的实例上。虽然这种做法会有些影响，但是对于正面案例的研究会取得更为有益的结果，会获得更大的改进。这就是为什么美国建筑业协会和澳大利亚建筑业协会制定的原理文件在这方面迈进了重要的一步的原因。通过制定出一套清晰的概念性提纲，可以将那些体现包含好的可建设性原则的优秀实例加以识别和应用到新的项目中去。

已经有人提出应该建立一套可建设性成功指标索引，而且该索引要集中在成功的例子而不是失败的例子。其基本的假设是在楼宇采购周期中对可建设性信息的管理和决定过程的记录可以取得最大的收获。这并不是想争论施工技术发展的重要性，而是将可建设性看作是由管理来驱动的过程，而不是由技术来驱动的过程。与这一观点相类似，建筑研究基础(BRE—Building Research Establishment)评估出，近90%的设计错误来自于没有很好应用现有知识。因此加强了关于可建设性中最为重要的一个方面，也就是缺乏对信息的管理的争论，而不是缺乏信息。当然，在技术领域也能够并且已经取得了一些收获。一些实例，特别是在英国包括模块式协调，设计理性化，标准化，预制构件和干作业等的使用。然而，如果在项目决策过程中没有适当的工具来管理信息，没有一个清晰的概念性模型作为操作参考，要想完全发挥出可建设性方法的潜能是不太可能的。

正像上文中提到的，以前为制定可建设性指标或规模所付出的努力都用在案例研究中，而且这些案例研究似乎将注意力集中在项目的失败的方面。更为有效的向前的方法是将可建设性的目标集中定义在决策过程上。

基本上，可建设性可以被归结成为有以下特征的项目要素：
- 在那些决定项目过程的人们的影响范围内
- 可通过已定成功指标来衡量(反映了施工的难易程度和项目的竣工质量)

从这一定义来看，可建设性具有三维性。它们是：
- 参与者(干系人和决策者)
- 可建设性的所有因素
- 楼宇采购过程中的各个阶段

图6.6表现了可建设性的三维概念模型。这三维模型定义了可以提出可能的解决方案来提高可建设性的空间范围。这个三维模型提供了一个对每一个项目都需要考虑的主要因素来提高可建设性的框架。

参与者

参与者包括两个小组。第一个组是干系人。也就是那些对于项目过程的可能产生的结果很感兴趣的人。第二个组包括那些其做出的决定会影响这些结果的决策者。其中一些可能出现的结果会直接或间接与项目的可建设性有联系。

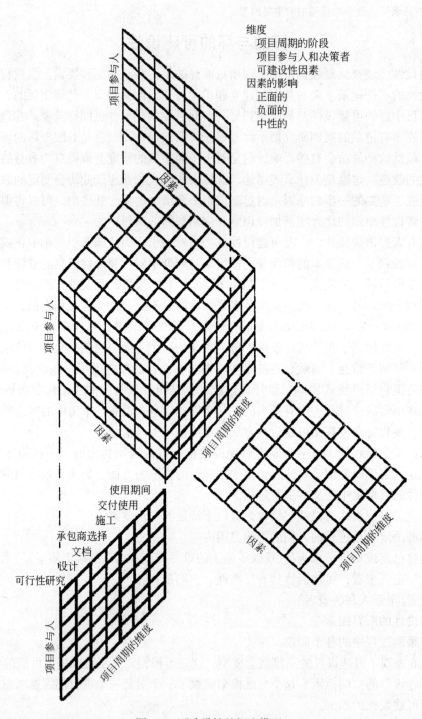

图 6.6 可建设性的概念模型

可建设性因素

这些因素(大体上等同于美国建筑业协会的方法的原理)是指,就决策而言,"参与者"

可以对其产生影响或回应的因素，以及那些会对施工的难易程度或项目的竣工质量产生影响的因素。这些可建设性因素可能是外因，内因或者是项目的最终目标(而非可建设性的目标(参见图 6.2))。

项目生命周期中的阶段

可建设性并不局限于设计连施工的楼宇采购方式中。对可建设性产生影响的决策包括那些在项目设计阶段之前所做出的决定。与此同时，对可建设性的评定的标准还包括了施工后的楼宇的表现。

应 用

在一些情况下，一种对于一个项目的可建设性会产生积极的影响的因素，对于另一个项目的可建设性可能会是中性甚至产生消极的影响。在这些约束之下，项目决策的目的也许就是降低或削弱负面影响和扩大正面影响。(这一工作仍处于发展阶段，并与决策支持系统中的普遍的特性的研究相联。)

实现可建设性的利益

虽然帕累托(Pareto)原则指出在项目过程中越早考虑到可建设性思想，其产生的影响将越大，而且对于时间与支出的节省及质量的提高也越大，可建设性的利益在项目采购过程中的各个阶段均可发生。据称，可建设性管理方法的实施能导致项目在时间、造价及质量上带来量化的明显改善。除此之外，可建设性管理方法还可导致项目过程和最终成品的质化上的提高。这些优势包括：
- 更好的项目团队合作
- 得到改进的工业关系
- 更为进步的培训
- 更高的生产率及更顺利的现场操作

1988 年的一项对可建设性在某一项目管理中的应用进行调查研究的总结说：
- 施工人员对于项目的设计的贡献是非常重大的；
- 项目的不同阶段中施工与设计间的反复关系带来明显的利益，包括节省成本和时间，简化施工难度，消除了工会界线及其他工业关系问题等；
- 设计的合理性，包括设计简单化、模块化及细部设计的重复性，对于可建设性的实现至关重要；
- 可建设性的实现受建造过程中诸如施工技术(系统、项目计划及进度安排等技术因素的影响；
- 还有许多其他因素，特别是非技术性的因素，而与楼宇项目管理(例如项目采购方式，交流和协调，质量管理)有关的，同样需要作为实现可建设性的组成部分看待。

美国建筑业协会，澳大利亚建筑业协会和一些其他组织都记录了一些体现可建设性优越性的实例和详细设计情况。对于那些希望对可建设性进行深层次研究的读者们，我们推荐亚当(Adams)和弗格森(Fegson)二人的文献，他们曾列举了大量的运用了可建设性原则的细部设计及施工现场组织的实例，在这些实例中，亚当(Adams)的文献罗列了 16 条设

计原则,并且均配以实例说明。以下列出这16条设计原则:
1. 全面深入进行调研;
2. 在设计阶段考虑可达到性;
3. 在设计阶段考虑材料贮藏问题;
4. 设计应考虑在地面以下的施工时间应为最短;
5. 设计应考虑尽早将楼宇围封;
6. 采用适当的材料;
7. 设计应考虑现有可行的施工技能;
8. 设计应考虑装配要简单;
9. 设计要尽可能促进模块化和标准化施工;
10. 最大限度的使用机械设备;
11. 允许范围内的误差;
12. 允许实际可行的操作顺序和方法;
13. 避免返工;
14. 制定计划时考虑避免下一道工序损坏前一道工序的情况;
15. 设计时考虑安全施工;
16. 清楚明了的互相交流。

其他文献给可建设性的利益做了大体描述。由澳大利亚建筑业协会记录在案的案例分析对在可建设性管理的系统性实施过程中实现的表现进行了典型的说明。其中的两个实例,在维多利亚州的丰田汽车制造厂和在南澳州技术园区的澳大利亚传媒中心两个项目,给出了对于实施可建设性所带来的利益的味道:

丰田汽车制造厂包括一幢面积为 120 000m^2 的楼宇,并且该楼宇在 107 个星期以内建成(比原计划提前 3 星期竣工),成本为 1 亿 6120 万澳元(比原预算低 1880 万澳元),并达到了很高的质量和安全标准。项目中采取的重要举措之一就是建立一套主要的屋顶和墙面模型,来检验并建立连接处和开放处的细部构造,来达到油漆工作间所要求的空气密闭性和气候条件状况的要求。

澳大利亚传媒中心需要建造一座面积为 7300m^2 的建筑设施,并且在 9 个月的紧密时间计划中完工,成本为 1200 万澳元,还有成本节省,这些节省使得在预算内增加附加规模的项目成为可能。该工程所实施的可建设性管理策略确保了项目团队成员的密切协调和联络,以及尽早解决设计,细部设计和施工方法等问题。

将可建设性的利益量化

计算并量化可建设性的利益并不能直接实现,特别是象在本书章节中,将可建设性作为一个广义的概念应用的情况下。这其中是有一定的原因的。首先,许多可建设性的利益,例如较好的团队工作,具有质化特性而非量化特性。第二,从可建设性操作行为角度看,经常会产生协同作用或者撞击效应。换句话说,整体通常大于各部分的总和,因此对于可建设性方面的时间,成本和质量改进上进行的简单的计算及累计不能保证已经掌握了可建设性带来的全部影响。例如,如果当地政府规划部门的代表被包括在项目团队当中(正像澳大利亚传媒中心项目这一案例),这也许会导致边审批边施工,因而可能会对工作

第六章 可建设性(Constructability)

流程产生根本性的影响。第三，为什么可建设性的利益很难被量化的原因是个方法论的问题。计算可建设性利益的理想方法是采用了"可建设性"的项目和没有采用"可建设性"的同类型的项目即"非可建设性项目"间的比较。非常明显这种方法是不可能和不可行的，并且，即便可能，也只可用于非常有限的几个例子。可以发现这样一些实例，即大公司的高级行政人员准备在公众场合宣布："他们已经在三个主要项目上进行了可建设性试验，并节约了5%的成本和13%的时间。"这些声明的真实性是不需怀疑的，然而这种用量化的方法从基本上就值得质疑，因为其比较是基于假设的基础上，它假设出了一种可建设性根本未被使用，或者使用效果不佳的情况来进行比较。

目前已经存在着一些在时间、成本和质量上将管理行为和楼宇项目表现联系起来的想法。

"……可建设性分析的实践工作在澳大利亚要比在美国或英国更为普遍……，对于一幢20~30层高的楼在澳大利亚设计过程需要用去18个人/月的时间，而在美国只需要4个人/月的时间。"

首次试图对可建设性进行严格量化的更进一步的工作总结说："大型项目的施工时间的变化的系数几乎是4:1；工期变化的成本含义是指贷款利息和其他租赁的节省以及尽早将楼宇租出去所带来的利润，将这些节省的成本与平均的工期相比，节省可以达到楼宇成本的50%"。

虽然从类似的研究和从美国的经验中可以清楚的看到，可建设性的利益是很明确的和重要的，但是，一套用于标明成功的量化方法仍未实现。

结 论

也许有人会认为可建设性的优点已是显而易见，并且其原理与优秀的多专业团队协作工作的原理相差无几。这是一个合理的推测，且很难对其质疑。可建设性总体上是对资源进行调配，使之发挥其最佳效用。这样做意味着在团队成员间建立无间隙交流。因此，这意味着打破传统的桎梏，改变专业的思维模式。施工人员必须理解建筑师的观点，反之亦然。业主必须随时准备做出负责任的决定。项目团队的所有成员必须随时准备主动发挥作用，而且着眼于从立项到使用的整个楼宇周期。从此表述中，我们可以看出为什么可建设性仍不是一种普及的施工管理手段。佐特伯(Jortberg)曾将现有形式悲惨地描述为"设计者和工程师们不知道他们不懂什么"。从早期的英国建筑研究及信息协会的定义，"建筑设计在满足完工后楼宇必须具备的要求的前提下，使施工的容易。"至今，可建设性已经历经了很长一段发展过程，但是其前面仍有一段长路要走，来实现澳大利亚建筑业协会关于可建设性的目标，即"一套实现施工知识在整个建设过程中的整合，并使各类项目与自然环境条件限制达成平衡，以使项目目标与楼宇的可使用性得到最大体现的体系。"

第七章

全面质量管理

导 言

最近的泰晤士报纸(Times)中有一篇文章报道，一幢房子的买主反映，在他们搬入这幢房子后，仍然花了一个全国性的承建商103天才完成厨房的施工。另外，在这短短的时间当中，他们发现该房子有112处问题。这对夫妻花了82 000英镑来购买这幢房子。且不说这个，这幢房子在成交的日期并没有完工，油漆工在他们搬入的那天仍然在工作。大量的质量问题使得承包商的工作人员在之后的五个月里，每星期光顾此房子三四次。在报道中房东承认承包商已尽其所能来修补这些缺陷，但问题仍然不断出现。

当然以上举的是一个比较极端的例子，而且单独来看这也许不是主要的问题。但是这篇文章真正值得注意的是房屋建造协会给准备购买房屋者的忠告，内容包括：

- 通过律师或造价工程师来买。
- 注意买房合同，因为追索权将依赖于它的具体条款为依据。
- 除仔细阅读合同之外，买主应在律师的协助下帮助编写买房合同。

根据这篇文章，国家房屋建造委员会，也就是制订房子建造标准和规则的单位，同样也提出他们的建议：

- 检查承建商是否在国家房屋建造委员会注册。
- 从以前的买主那里了解该承建商的质量标准。
- 检查工地现场是否清洁和管理有序。
- 不时地到工地现场去查看。
- 一个好的标志是工地经理曾经在该类工作中获得过房屋建造委员会的奖励。
- 在房子的外边查找砖墙、屋顶、喷涂、管道和排水设施等方面的问题。
- 当房子有质量问题时，如果承建商还在从事本行业工作的话，第一个要找的就是他。

作者认为，上述建议反映了建筑业的建筑质量低劣的问题。假设这两个组织(房屋建造协会和房屋建造委员会)是要维护房屋买主的利益，他们期望房屋买主普遍能够分辨清洁及管理有序的工地和脏乱差以及欠缺管理的工地之间的区别。我们的猜想是：在一般人认为，所有建筑工地都是脏、乱、差的。还有，为什么买主要通过造价工程师和律师来买房子？文中的全国性的承建商已经从事建造和销售房屋几十年了。为什么他们自己不可以卖房子？

现在让我们来假设这幢房子是一辆汽车。想象一下，如果你花了几万英镑买了一辆汽车。当汽车油漆还没干时，制造商是否会让你把车开走？当你买一辆新车时，你是否需要找律师来帮你解释或重写购车合同？你是否需要不预期地到工厂去检查现场是否整洁和管理有序？你是否有能力来分辨整洁的和肮脏的汽车制造厂？而且，如果该车有112个缺陷，你会怎么办？你会让汽车制造商连续五个月每周到你家三四次来修理这辆车吗？不大可能。很可能的是你最终会要求换一辆新车，而且汽车制造商会认识到问题的不可接受而满足你的要求。

建筑业的所共有的问题之一是质量低劣的文化习惯。上述两个国家级机构的忠告也反映出了这个问题。似乎整个建筑行业觉得：在质量问题上，顾客要自己处理，质量问题并不是承建商或咨询顾问公司的工作。

毫无疑问，近几年来，建筑行业在工程质量有了很大提高，但是这种提高只体现在施工过程中的几个方面。工地的管理和房屋的设计也许是比十年前有所提高，但最后交付使用的房屋的质量还是不够好。这是因为建筑业不是真正的以顾客为中心。建筑业并没有把质量看作是满足顾客需要的一个总体问题，而只是把它作为处理有关设计、材料和工地安全的一系列步骤而已。一个以顾客为中心的全国性的建造商怎么可能会让买主花 82 000 英镑搬进一幢正在油漆和粉刷的房子呢？

没有以顾客为中心的做法，并不仅仅存在于建筑业，制造业也是如此。在来自日本产品的竞争下很多制造业因质量低劣而在竞争中付出失去市场的代价。很多这些企业已经认识到真正优秀的质量并不是靠对产品一系列检查的标准得来的，而是来自于以顾客为中心的全面质量管理方法。服务业在这方面尤其成功。全国性的超级市场就是在全面质量管理方面一个很好的例子。现在有些超市在父母在采购的时候，给他们的孩子提供"星期天学校"。

这就是全面质量管理的精华所在。这是以顾客为中心并涉及到组织的各个部门和人的全面质量管理方法。

全面质量管理的定义

在本书所概述的所有技术来说，全面质量管理涉及的范围最广。从某些方面说，全面质量管理就像一把大雨伞，把所有其他的概念都包含在它之下。也可能是因为这个原因，没有一个能被广泛接受的全面质量管理的定义。例如，林板斯（Rampsey）和罗伯特斯（Roberts）定义全面质量管理为：

"一个以人为中心的管理系统，其目标是在不断提高顾客满意度的同时不断地降低成本。全面质量管理是一个全面系统方法（并不是一个分开的方面或程序），是一个高层次策略的组成部分。它是横向贯穿于职能和部门，从上到下，包括所有雇员，而且前后延伸而包括供应商链和顾客链。"

将这个定义和下面的定义比较：

"全面质量管理是企业为了不断提高商品质量和服务质量所使用的所有职能和过程的整合，其最终目标是使顾客满意。"

事实上，这些定义很相似，都包含了全面质量管理的基本原理。第一，全面质量管理必须是一个全面的解决质量问题的方法。在过去，质量只在企业的一些部门得到考虑，比如最终产品或者顾客关系。全面质量管理关心的是将整个系统作为一个综合体单位。第二，全面质量管理是连续的。过去质量被视为一个对产品或者公司的某些方面有所帮助的系统。而全面质量管理应是一个连续的过程。现在的观点认为不论一个系统有多好，它还总是可以进一步提高的。最后，全面质量管理的目标是顾客满意。在过去，质量系统以提高产品质量为目标，而不是以顾客满意为目标。这两个目标不一定是相同的，因为有可能是将产品质量提高了，但却没有意识到这并不是顾客想要的。

图 7.1 概括了全面质量管理的三个驱动器(组成部分)：整合、以顾客为中心和不断提高。

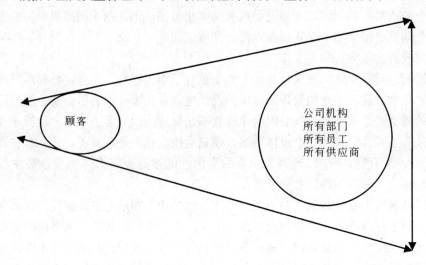

图 7.1　全面质量管理的三个驱动器：整合、以顾客为中心和不断提高

在细致检验这三个驱动器的细则之前，值得考虑和强调支持他们的几个概念：第一是检验什么是质量；第二是全面质量管理的发展历史；第三则是企业文化的概念。

什么是质量？

就像全面质量管理的定义一样，质量也有很多不同的定义。

以制造为基础的定义视质量为满足要求和规格标准的能力。这种质量的衡量是客观的，因为它纯粹是建立在某一产品或服务的能力能否满足一个预先制订的规格或标准的基础上。例如，我们可以衡量一个电热器是否散发出适量的热量，或者一个项目的施工进度是否按时完成。这种类型的质量定义的问题是定义没有表明所衡量的内容就是顾客需要的。这是一种向内着眼的质量衡量，而不能被认为是全面质量方法。

以产品为基础的质量的定义同样是客观的，它基于对产品的一个具体特征的衡量，例如，耐久性或维修，一个塑料窗的质量可以说比木窗的质量好，因为它使用期较长，而且所需维修较少。

以使用者为基础的质量的定义，却是主观的，其质量评价是基于产品对使用者的满意程度。这个使用者满意的概念是和价值管理一章中的功能分析的概念紧密结合的。使用者质量可以被描述为满足使用者需求的质量。而对建造而言，就意味着一个较低的产品质量可能具有较高的使用者质量。例如，对一栋仓库的建造来说，使用上述提及的塑料窗可以有较好的产品质量。但是如果该仓库的设计使用年限仅是 20 年，那么木窗可能有更高的使用者质量，因为木窗更能满足使用者提供使用期为 20 年的窗户的需求。

以价值为基础的质量的定义尽管包含一个上述的质量衡量，但以成本造价为内容。因此，从以价值为基础的对塑料窗和木窗的质量检验中，可以认识到塑料产品是更耐久和更易维修的，但也会被认为需要较高的成本。这个对质量的"最好购买"方法经常被消费者杂志采用。

因此，现代建设公司该采纳上述哪种质量定义呢？答案是所有这些定义都有一些作用。一个好的建设项目将会符合设计细则，满足使用者的质量标准和具有理想的价格。这听起来好像很抽象，但是这些各种各样的质量定义可以被阐述成一个更客观的主要质量尺度的细表。

使用功能

这是拥有该项目和它具备的那些主要特点的首要原因。对一栋医院的楼宇来说，也许就是病房、候诊室和手术室。

可靠性

指楼宇是否可以使用合理的一段时间而不出现问题。

符合性

符合具体要求的程度。

耐久性

楼宇在需要被拆毁之前所经历的时间。

服务性

楼宇在建好后所能提供的服务，特别对维修而言。

美观性

这是指楼宇的外形和给人的感觉。

质感性

这是指从楼宇的形象所得出的对该建筑物质量的主观判断。

表 7.1 进一步检验这些质量参数，看看一般的楼宇对这些要求的符合程度：

表 7.1 主要质量参数

质量参数	建筑的表现
使用功能	楼宇的绝大部分建筑达到主要目的没有？
可靠性	他们可靠吗？
符合性	他们符合具体要求吗？
耐久性	他们的使用期比要求的长还是短？
服务性	他们所需维修是否快而且维修后能有一个好质量？
美观性	他们是否内外都美观舒服？
质感性	使用者和顾客感觉这是一栋质量好的楼宇吗？

因而,"质量"一词很难定义,它存在于各种不同层面,它的范围可以从楼宇的各部分功能是否满足具体要求到整幢楼宇是否满足客户需要。可以独立的来判断质量,也可以连同其他方面衡量,比如成本。对一栋楼宇质量的评估可以基于上述七个主要参数来进行。要达到这些主要参数的高要求的方法是通过对所有建造过程的管理。这就是全面质量管理的本质。

全面质量管理的发展历史

质量并非是新的概念,它起源于制造业的以检测为基础的系统。为了减少有缺陷的产品流向市场,在制造过程中,产品需要检测。受检产品与标准进行比较,被清除出来的不合格的产品,或被废弃,或做修理,或者作次等品卖出。这种以检测为基础的质量系统有如下几个缺点:

- 没有检查出来的有问题的产品将流向市场和顾客。
- 这个系统的操作昂贵,因为这些操作基于纠正缺陷。
- 这个系统将责任从工人的身上转移到质检员身上。
- 这个系统不能表明产品不合格的理由。

由于以上这些理由,加之产品构造也越来越复杂,以检测为基础的系统已被统计抽样的质量控制系统所取代。这类系统的先驱之一是丹明(Deming),丹明(Deming)的工作致力于通过减少设计和生产的变更来提高质量。丹明(Deming)认为变更是导致产品质量差的最主要原因。他相信变更来自于两个方面:普通原因和特殊原因。普通原因是由于生产过程的问题而导致的,而特殊原因来自于某一种或成批的材料。对建筑业而言,普通原因可能包括:浴盆和釉砖的交界处。因为这也许是设计本身的问题,而不是材料的问题或工人的问题。另一方面,一个特殊原因可能是砌砖工人混合的砂浆太稀了。

为了通过减少多样化而提高质量,丹明(Deming)列出了一套十四点管理系统。这些内容着重于过程,因为丹明(Deming)认为导致变更的原因是系统本身,而非工人。他的内容包括:

1. 建立并发布公司的目标;
2. 学习新的质量哲学;
3. 不依赖大规模检查;
4. 不要仅以价格作为判断工作好坏的单一标准;
5. 不断改进系统;
6. 公司培训;
7. 公司领导;
8. 祛除担心,建立信任;
9. 消除部门之间的障碍;
10. 消除口号和工作指标;
11. 消除数量配额;
12. 消除隔阂,增加工作的自豪感;
13. 制定公司自我提高和培训及再培训计划;
14. 采取行动来推动改变。

丹明(Deming)认为一旦一个质量系统建立并运行起来，将带来质量链的系列反应，如图 7.2 所示。也就是由于质量提高而带来成本下降，而且误差和拖延都会减少。因此生产率和市场占有率都提高了。又因此带来更好的质量和更低的价格，这意味着公司有了更强的竞争力，并提供更多的就业机会。

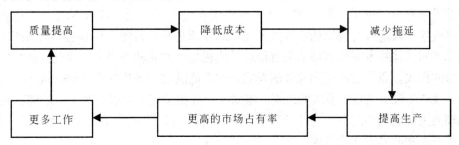

图 7.2　质量链

与丹明(Deming)同期的朱然(Juran)，另一位美国质量顾问，将质量控制技术介绍给日本人。由于近三十年来要求持续上涨，日本产品逐渐占据了西方市场。原因主要在于日本出众的质量体系使他们的产品质量远高于西方同类产品。当质量管理在日本已经以自己的方式发展成一个独立科目时，西方管理方法却一直停留在 20 世纪 50 年代的初步阶段而没有改变。

毫无疑问美国产品和日本产品质量的差距是美国和欧洲质量革命的动力。革命一词不是随便用用而已，最近几年已经见证了对质量认识的根本性改变。在 20 世纪 80 年代以前，人们将质量只是看作内部的问题。现在人们认为质量是由客户或是外部因素来驱动的。人们认为质量是和一个观念紧密相连的，这个观念就是一个机构是由一系列的过程组成，每一个过程都有相应的客户。这些客户可能是机构内部的。但通过尽量满足他们，最终产品的质量将会被提高，用户也会满意。另外，通过以上对质量的定义来看，和传统的质量控制技术相比，全面质量管理是一个非常广泛的哲学观点。它包含组织的方方面面，从高级管理层到客户以及供应商。最后，以前的质量控制技术依赖于检查，而全面质量管理强调的是预防。

如果认为丹明(Deming)是质量管理惟一的领导，那么会有些不妥，也有一些实施这理念的人并不赞成他的方法。例如，朱然(Juran)关于质量的观念并不像丹明那么广泛。他不想从根本上改变，而是在组织原有系统的基础上提高质量。克罗斯比(Crosby)关于质量哲学的观点又不一样了，他非常强调行为方面，不像丹明用统计分析。

因此本章并不仅仅以丹明的理论为基础，而是综合了许多可用于建筑业的思想。我们不是在建议一个固定的全面质量管理过程，因为没有这样一个东西存在。全面质量管理是一套哲学理论而不是一种技术。

一套哲学理论暗示了一种思维方式，从这个意义上讲，全面质量管理提供了一个最大的挑战。人们的思维方式是由文化决定的，为了在企业中采用全面质量管理，该企业必须经过一次文化转变。这种文化转变的想法和建设管理中的一个现代观念——模式转变很一致。

模式转变的需要

当一个企业正在发展时，它以现存的体制范围运作。任何改变和发展也都在系统界限之内。系统的界限是有用的，因为它说明了其运作中的制度和原则已被接受和认可。然而这些分界线总有一天过于局限，因而为了解释正在发生的变化，或者满足一些发展的需要，调整这些界限还是有必要的。这些界线的改变就是模式转变。在本书中提到的很多全面质量管理和其他技术的作者都认为目前建筑业正处于模式转变当中，而且是新的模式正在代替旧的模式。我们已认识到现存的模式已经不能满足全球化的竞争的建筑业，因而必须改变。表7.2揭示了新旧模式的区别。本章在一开始概述了以客户为中心的课题是全面质量管理的关键驱动器之一，表7.2表示以客户为中心的模式，并被分解为质量、度量、定位、干系人和产品设计几方面。

表7.2 新旧模式

主题	旧模式	新模式
质量	符合具体要求	顾客价值
度量	内部效率测量	与顾客价值相关联
定位	竞争	顾客环节
关键干系人	干系人	顾客
产品设计	内部销售自己建造的商品	外部建造顾客所需的商品

一个运行在老的模式中的企业将是一个追随者，其首要任务是使干系人和高层管理盈利。而且与其他企业的成功相比较来判断自己的成功程度。对其效益的评判也是与其预先制订的指标作对比，而对造成这些指标的原因却很少研究，而是通过检查对质量实现控制。最后，该企业认为质量主要是满足具体性能要求。相反一个在新的模式下运行的企业找出顾客的需要然后生产出满足这些需要的商品。质量以及对效益的检测都与顾客价值相连。市场定位是以市场占有率为出发点的。最后，短期盈利并不是目的，因为被看作主要干系人的人是顾客而不是股东。

本章的导言中概述了全面质量管理的三个驱动器是：企业的整合，以顾客为中心和不断提高。表7.2只提及了新模式的一个方面，即以客户为中心。另外，新模式也体现整合和不断提高。在整合方面，企业各部门的功能之间的障碍正在消失，而且人力和组织行为在组织中正变得更加重要。过去高大的等级制度结构正在被强调团队工作的平等结构所代替。从不断提高的角度讲，也在从自上而下的短期管理模式向更加开放的，注重长期计划的，由顾客驱动的运行模式转变。

建筑业的文化转变

模式是指共有知识体的特征的思维和行为方式。文化是支撑着这些思想和行为的社会结构。要成功实行全面质量管理，就要改变模式。为了改变模式，就需要转变形成建筑业的企业文化的转变。这些所需要进行的文化转变可概括总结为表7.3所示。

表 7.3 全面质量管理所需的文化转变

由	到
满足具体要求	不断提高
按时完成	满足顾客
着重于最终产品	着重于过程
短期目标	长期目标
以检查为基础的质量体系	以预防为基础的质量体系
人作为成本负担	人作为资产
最低成本供应商	质量供应商
划分的组织	一体化
自上至下管理	雇员参与

不过，企业文化是社会文化的反映，这意味着它很难改变，因为它是存在于一个使社会成为一个整体的规则之中的。但引进新的企业文化到工作环境中也不是不可能的。一个在英国开展业务的日本公司的成功的例子便是这方面的一个很好的证据。这个公司成功地改变了自己的企业文化，使之和一个完全不同的社会系统相适应。

一家日本公司在英国新开一间工厂，改变其固有的企业文化也许会比一家英国建造公司从一种文化向另一种文化转变要容易。不过，意识到改变企业文化的可能性，而且这是唯一能实现全面质量管理的途径可以是一个更大的改变的驱动器。此外还有几个改变企业文化的方法，改变可以通过如下几个方法来达到：

- 管理的角色
- 培训和提高
- 过程和系统
- 员工参与

仔细观察一下其中的一个方法，传统意义的管理角色是：

- 计划
- 组织
- 命令
- 协调
- 控制

现在这个角色需要包括：

- 名义领导
- 领导
- 联络
- 监督
- 宣传者
- 发言人
- 企业家
- 处理干扰者

- 资源调配者
- 谈判者

这是一个融入企业机构中的更广泛的角色，它更关注人，是对现代管理的更现实的评价。想要采用全面质量管理的企业必须认识到不仅改变文化是需要的，而且必须认识到这个文化改变不像人们通常认为的那样是不可能的。表7.4概括了一个改变文化的主动方法。此表也举例说明了如何在建筑业中实行这项策略。

表7.4 一个主动的文化改变的方法

策　略	实　例
检验目前状况	差的分包商关系
确定所期望的价值系统	良好的分包商关系
建立政策来实现价值系统	持续使用同一组分包商 建立内部调解程序来处理分包商的不满情绪 在雇佣新的分包商之前先要检查他们已经完成的项目业绩和水平 采用公平的工作章程
全面质量策略	对分包商进行分组 设计调解程序 拟定公平的工作章程
实施全面质量管理	将新策略传达给分包商 就新方法和新政策培训雇员

本章的第一部分曾分析了一些支撑全面质量管理的理念，不仅是质量的概念，还有建筑业文化转变的需要。本章下一部分集中于全面质量管理的三个主要驱动器，以客户为中心，整合和不断提高。

以顾客为中心

以客户为中心的思路之所以重要是因为它给管理者提供了一种生产能满足客户需要的产品和服务的思维方式。对于建筑业来说，其问题是很难定义谁是顾客。从全面质量管理的角度看，存在两种顾客，即最终用户和内部顾客。然而在建筑业，最终用户也不容易定义，因为经常会有很多最终用户，而且每一个用户的需求都相互冲突。而价值管理技术是为了综合考虑一个项目的所有最终用户的需求。虽然这还是个新观念，作者仍建议将价值管理与全面质量管理计划相结合。因为以我们的观点，这种技术相对其他所有技术而言，更适合于将建设项目中出现的各种利益冲突合理化的复杂过程。

内部顾客这个理念已经在前面特别是目标管理一章里分析过了。内部顾客的理念是和商业过程紧密相连的。就像目标管理和重组一样，全面质量管理是着重于过程的。因此，内部顾客是过程的顾客，或是在这个生产过程链上的下一个人。怎样提高对内部顾客的服务将在下一节"整合"中分析。

整　合

传统企业被分成很多部门，每一部门行使一个功能，比如预结算部门。在这些功能部门当中也划分了级别。人们已普遍接受企业这样划分的作法，尽管这种划分方法有如下一些问题：

- 形成了各自为政的帝国
- 交往范围窄
- 价值标准分化
- 部门之间竞争
- 部门间的首要任务互相冲突
- 僵硬的结构

从另一方面，整合的企业有单一的目标和共同的文化。交流增加而且重视对个人的尊敬而不是对他们所在的工作部门的尊敬。全面质量管理的主要目的之一就是消除企业内部分化，使机构向更加整体化发展。这个目标可由如下几种方法达到：

过程管理

如上面所解释的，一个企业有内部顾客，也就是商业过程的顾客。这一商业过程的概念是全面质量管理思想的核心，因为它相信如果满足了所有内部顾客，那么这些为外部最终用户提供的产品或服务也将被提高。对一个建设项目来说，当中通常要涉及好几家企业，这通常是产生管理问题的原因之一。然而，商业过程的理念跨越了这些界限。因为这种理念认为所有公司的所有人都是顾客。如果主承包商视分承包商为顾客，反之亦然，可以想象建设过程将进行得顺利多了，并也将给业主和最终用户一个更好的项目和服务。尽管在建筑业可能持有相反的观点，但建筑业与制造业没有不同，也就是不同的公司生产出组成最终产品的部件。制造业与建筑业相比在很大程度上更愿意接受这样的情形，即最终产品的总体质量是参与生产过程的原材料及生产状况的反映，并不是公司本身。

从全面质量管理的角度来讲，顾客被定义为是任何一位某项工作或活动行为的受益者。顾客可以被划分为：

- 内部顾客，即对过程而言的位于组织内部的顾客。（比如销售部门）
- 外部顾客，即对过程而言的位于组织外部的顾客。（比如分包商）
- 最终用户，他们出资并拥有最终产品。（比如业主）

一般假设是通过改进过程来满足内部和外部顾客的需要，将使最终产品的质量标准得以提高。其内在的理念是质量控制不在于最终成品的检测，而在于每一个过程的控制。这是和西方的传统管理模式不一样的。西方的传统管理模式倾向于检验最终成品，西方质量管理系统一般来说是自上而下的，而且倾向于由实施这个系统和等待产品结果来衡量成功与失败的管理者来设计。这种方法可以被描述为单一的思维：即决定完全取决于对最终产品的计量检测。计量检验通常有一些经济参考指标。

单一思维方式的影响

单一思维偏重于数字，是一种很差的检测企业的方法，而且可能会导致：

停　滞

如果一家企业总是用同一个目标作为衡量成功的标准的话，毫无疑问这个企业将是僵滞的。企业所存在的环境是不断变化的。因而保持检验成功的标准一成不变是毫无意义的。

削弱质量基础

质量不能单独用计量的方式衡量。因为依赖计量来检测成功无疑将导致注意力从质量转移到数量达标上。

缩短时间

如果以数量作为目标就存在着只关心时间的倾向，只从季报表或者中期报告来评估成功程度，会鼓励员工将工作的任务看作是一系列的短期行为。这种短期行为主义是对质量和企业长期的健康运转是有害的。

抑制投资

把注意力放在短期结果和目标将削弱长期投资。

各自为政的行为

这是大企业的普遍问题，而且出现于当员工只认识自己的部门和自己部门的活动，而不是把企业当成一个整体的时候。它导致了部门之间内部斗争而且缺乏对整家公司承担义务的局面。

过程思考

与单一思维相反的是过程思考。这种思考并不集中于对最终成品的计量检测，而是集中在过程上。过程思考背后的观点是导致成功和失败的原因在于造成这种结果的系统，最终原因在于构成这种系统的具体运作过程。

延长时间标尺

过程提高在本质上是不间断的。因此一个以过程为出发点的企业从长远打算，更倾向于鼓励投资和获得长期成功。

创建企业文化

前面已经说过，全面质量管理的基础是转变企业内部的文化。致力于过程有助于达成这种文化转变，因为这种方法将焦点从最终成品，短期结果和各自为政移开。

倒转的组织

倒转的组织也可以作为整合企业的一种方法。倒转的组织将那些提供服务的人员作为组织结构的顶层，而非经理，因为他们来自第一线。

员工参与

另一个实现企业整合的方法就是通过员工的参与。所有员工都是企业的一部分的理念，不是一个简单的不切实际的想法。其目标是通过对企业员工授权参与决策和解决问题来改进体制。其目的是鼓励员工参与到过程中和增加他们的主人翁精神。实现员工参与可以通过以下几种方式：

质量圈

质量圈把员工组织到一起，参与鉴别、分析和解决与某一过程相关的问题。质量循环是很重要的，因为它鼓励主人翁精神，因而提高过程。

质量圈可以是主动和被动两种。当某一过程出现具体问题时，则采用被动圈。而另一方面，主动圈是永久性的团队。这两者中，主动质量圈被认为比较好，因为在正常状态下比出现问题时的过程学到更多东西。

质量圈的要素

- 人
- 技能
- 时间
- 地点
- 资源

质量圈组织的阶段

- 识别——识别一个有问题或者成为主动质量圈的课题的过程。
- 评估——识别某一过程的所有问题和潜在的问题。区分可以被纠正的问题和公司不能控制的问题。
- 解决办法——找出一个解决问题的最优方法。
- 汇报——向高层管理汇报质量圈的结果并提出一个改进的系统。
- 实施——实施被改进的过程。不过由于过程改进是一个持续的过程，所以必须跟踪新的过程，而且如果需要的话要继续改进。

质量圈在解决直接关系到劳动力的问题上会特别有用，因为这些问题有实践的本质。但是在更广的范围会有一些问题，因为在这样的范围中，中间管理层认为质量圈对他们权力构成威胁。

跨部门的团队

跨部门的团队通过横向运作，而没有遵循典型的等级结构，促进了企业的整合。这些团队的基础是通过它们运行的过程来协调企业各部门间的合作。

培训和发展

没有培训，就不能有员工参与。全面质量管理的培训通常分三类：增强质量宗旨，有关具体工作的技能培训和全面质量管理原则。全面质量管理关键的原则是解决问题。解决问题是管理的巨大领域，因此下面列出了一些主要的解决问题的技术。

暴雨式思考(Brainstorming)

暴雨式思考在价值管理一章中作为产生创造性设计方案的方法已经概述过了。暴雨式思考也可以作为一种解决问题的方法。就象价值管理一样，暴雨式思考也有它的规定。在开始暴雨式思考解决问题研讨会之前，除了一些完全中性的术语之外，不应该有关于问题的定义。也不应该有关于解决问题的建议。应该限制建议解决问题办法的时间，而且所有的建议方法都要记录下来，不管这些建议看起来是杂乱的。需要再次强调，应该鼓励不寻常的想法，因为经常是这些想法形成了有效可行的解决方法。

帕累托(Pareto)分析

帕累托(Pareto)分析包括计算问题发生的成本和频率，把它们按重要性来放置，以便显示出最高成本的问题，把它单独挑出进行改进。

因果图

这是由考路·爱思卡瓦(Kaoru Ishikawa)发明的，也被称为鱼骨图或爱思卡瓦(Ishikawa)图（图7.3）。

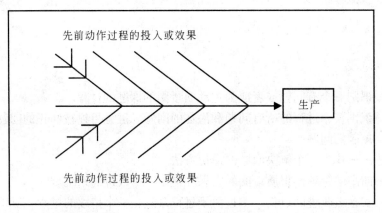

图7.3　埃斯卡瓦(Ishikawa)图

在目标管理一章中提到，一个过程有输入、过程和输出，而且一个企业又由成百上千个互相联系的过程组成。在鱼骨图中，一个原因是某一个过程的输入部分，而效果是过程所产生的结果。当一个过程运作正常或处于被控制状态时，此过程的结果或输出是可以预测的。当过程失控时，后果就不可预料。爱思卡瓦(Ishikawa)认为，对原因的调查是最关键的。他认为尽管有些东西看起来象是原因，但实际上它只是前一个过程的效果。通过沿着鱼骨图往回追踪，这个效果就可以找到。不过，只有我们知道什么是真正应该发生的效果时才可以找到该效果。基于此，当过程处于被控制状态时的研究和其处于失控状态时的研究是同等重要的。

表现评估

表现评估也是一个鼓励员工参与的方法。然而传统的评估方法不适合公司实行全面质量管理。表现评估的重点应该是针对公司的目标，而不是个人的贡献。

"再多一个"理论(The one more theory)

"再多一个"理论运作的基础是，企业内的每项工作都应该有至少两个人能够胜任，而企业内部的每个员工也应该能够完成他们日常工作以外的另一项工作。

全面质量管理的包容性

全面质量管理的包容性决定了公司应包括供应商和顾客在全面质量管理过程之内。在本书包含的所有管理系统中，也许全面质量管理比其他系统更可以归于软性系统。因此，不可能定义典型的全面质量管理系统是怎么样的，但可以说，最能体现公司企业通往全面质量管理的特征如下：
- 远见
- 关心顾客
- 失败的打击
- 以防为主
- 部门间零障碍
- 对竞争的跟踪
- 高层管理行动
- 高层次的培训

不断提高

全面质量管理的第三个驱动器是它是一个不断提高的过程。不断提高的实质是即使一个组织是盈利的，并且市场占有率也很高，仍然要继续寻找提高的途径。我们已经重复得不能再重复了，全面质量管理不是一个系统，而是一则哲学原理。其重要的一点就是没有哪一个系统是完美的，它们总是可以再改进。

质量成本和质量所造成的成本

读到这里，也许许多读者已经得出结论说由于规模或是利润率的原因他们的公司不可能实施全面质量管理。然而这种观点总是在一种误解之下产生的，这种误解就是关于质量的实际成本是什么。在根本上有如下三种关于质量成本的思考方式：

高质量意味着高成本

没有成本的增加，就不可能提高质量，而且公司从提高的质量中获得的利益弥补不了额外的成本。

用于提高质量的成本少于由此而获得的节省

第二种质量成本的观点认为，用于提高质量的额外成本少于重新建造或是废弃残次品所造成的损失。

第一次正确之方法

质量成本的这一观点比上述列举的两种更加广泛。这种观点认为，质量成本是那些超出如果产品在第一次生产时就没有质量问题的成本之上的那一部分开支。成本不是直接的开支，还包括失去客户，失去市场以及其他潜在成本。成本不是用被丢弃的或是重造的成本来衡量的，而是指那些相对于在第一次就生产出零缺陷的产品所需的成本而超额付出的实际成本。这是大多数全面质量管理的实施者的观点。

这些关于以零缺陷来衡量质量的成本的方法，可以归为四类：

预防——这是指消除或阻止缺陷发生的成本。例如质量计划和培训所需开支。

检查——这些是指在将产品运载到顾客那里之前，把劣质品检查识别出来所发生的成本。检查所带来的成本就是这样的例子之一。

内部失败——这些是生产过程中发生的成本，包括废弃和返工所带来的成本。

外部失败——这些是被拒绝或被退还的产品所造成的成本，包括一些隐藏成本，比如顾客不满意而导致失去市场份额。大多数全面质量管理实践者认为，质量成本陷入了冰山原理，即只有水面以上的一小部分是可见的，比如返工成本。然而隐藏在水面以下的不可见的部分，在成本中却占了更大比重，甚至有可能达到全部成本的 90%。

国际质量标准，例如 ISO 9000

ISO 9000 是五套世界范围的关于设立了质量管理的要求标准的其中一套。ISO 9000 不是产品标准，而是质量管理体系的标准。它在欧盟得到广泛使用。这套标准很普通因而应用于所有部门和所有行业。到 1992 年为止，英国已有 20 000 多家公司采用此标准并获得认证，而且欧盟以外的 20 000 多家非欧盟公司也已经注册。日本有关部门也力荐他们的公司进行注册。

申请 ISO 9000 所需的文件提供是非常繁琐的，而且需要编写质量手册，编写所有相关程序和所有相关的具体操作说明。

认证的好处

- 更大的顾客忠诚度
- 市场份额的提高
- 更高的股票价格
- 减少了买方打来电话要求服务的次数
- 更高价格
- 更高的生产率
- 成本降低

认证也有一些缺点。正如本章一直强调的，全面质量管理是一种哲学方法，因而人们认为严格的 ISO 9000 认证体系与此概念相反。它有与全面质量管理建立的目标相对抗的倾向，把质量的重点从客户转移到获取认证，并使获取认证变成了主要目标。与此类似，对质量的责任也从所有工人那里转移负责获取认证的质量部门。

对建筑业的研究表明，无论如何还没有确定性的证据来证明建筑业内认证的益处。然而我们必须切记，ISO 9000 这样的质量体系和全面质量管理并不是同一样东西。另外，对认证所作出的批评也许并不是针对执行这套体系本身，而是针对执行这套体系的方式。

改变管理

本章很明显地提出采用全面质量管理要涉及到公司各个层次的改变。改变不会自己有效的发生，它需要有人管理。管理改变就是控制改变的过程和把握改变的方向。改变的管理包括四个阶段：

- 建立改变的需要
- 获得和持续许诺
- 实施
- 回顾

有效管理改变的重要的一点就是必须包括公司的所有成员。这点和目前的管理观点是相反的。目前的管理倾向于由高级管理人员作决定进行改变，然后往下传达，而且很少解释理由。还有一种方法可选择，即蘑菇法，管理者们只是简单的把一些想法传达到某种他们自己都不很明白的环境中，然后等着看会发生什么事情。

即使改变已被管理有序了，也还会有一些障碍，包括成本，时间不足，员工的原有观念，行业文化和缺乏能力。识别和了解这些有关改变的障碍对于克服它们是很重要的。

全面质量管理的方法

任何关于质量和全面质量管理的书都会包含许多质量和全面质量管理的方法。这其中的大多数方法已经在制造业被广泛检验。然而，他们是否适合建筑业目前还不清楚。所以下面只是简单提及这些方法而不作详细介绍。如果读者希望作进一步了解，请参阅本章后面的参考文献。

- 大古奇(Taguchi)方法
- 失败模式和效果分析
- 统计过程控制
- 准时制

如何实施全面质量管理

要实施和维护全面质量管理系统，首先要有一个战略计划。此计划应包括：

质量的管理

- 质量定义
- 全面质量政策
- 全面质量战略
- 全面质量文化

人力的管理

- 所有员工应明白自己及其他人在组织中的位置

- 许诺
- 团队工作
- 教育
- 开放管理

过程的管理

- 过程设计
- 过程控制
- 过程提高

资源的管理

- 财富生成
- 质量的成本
- 资源保存
- 资源计划

凯审(Kaizen)

在这里有必要提到与全面质量管理有关系的一个日本概念凯审(Kaizen)。凯审(Kaizen)策略可能是日本管理中一个最重要的概念。凯审(Kaizen)策略认为任何一天都不可以虚度,公司里一定要在每一天在某些方面取得一定的进步。在日本,管理被认为有两个功能:保持和提高。凯审(Kaizen)意味着,现实中每一个微小的进步都是持续努力的结果。与全面质量管理不同的是,它运作于已有的文化背景内,不需要文化转变。凯审(Kaizen)这类系统在西方是否有效还不得而知。

目前对建筑业全面质量管理的研究

在全面质量管理领域,研究的一个部分就是认证体系。研究检验了 BS 5750/ISO 9000 的引入对与建筑业的有关企业所产生的影响,并强调对认证的利益存在着不同的观点。认证与建筑师的实际工作的关系也被检验,并且发现与建筑承包商相比,建筑师们实际工作中采用质量保证体系的进程非常缓慢。实际上获得质量保证系统认证的建筑师的实际工作只占全体的很小的百分比。

研究还表明,认证也许并不是最好的办法。建立全面质量文化和改进质量最有效的方法或许是公司发展自己的质量改进团队。

对南非承包商的质量控制的研究表明,承包商认识到质量系统的重要性,因为它是承包商获得投标资格的要求之一。然而在英国的研究表明,南非的承包商发现质量系统的形式对他们来讲是个问题,他们没有意识到可以有一个适合他们自己企业需要的质量系统。许多承包商都没有意识到,要使质量系统行之有效并不一定要将之正式化。

在另一个领域的研究是采购和质量管理的关系。这是一个新的课题,尽管承包商可能觉得设计连建造的采购方式是解决质量问题的最合适的方法,但还没有具有说服力的证据来说明这点。

结 论

最近，在作者的一次有关改进过程的讲座上一个学生提出，如果过程本身没有问题，为什么要试图改进它。在某种意义上说，这种想法是和全面质量管理的文化相违背的。全面质量管理的思想认为没有一个看上去很好的系统是不可以被改进的。本章反复强调，要有效实施全面质量管理，比本书提及的任何其它管理技术都要求改变文化，改变观念。如果意识不到这一点，完全不可能实现全面质量管理。因此本章没有详细介绍全面质量管理的操作过程，而是介绍成功实施全面质量管理所需要的文化环境。

全面质量管理从根本上来说是一种态度，即企业的主要驱动力是以顾客为中心。而这又是通过企业内部的整合和不断提高得以实现。这就是全面质量管理的三个基本要素，这其中有各种不同的系统，而这些系统之中又有各种不同的技术。有一些全面质量管理系统是高度公式化的。然而没有证据表明，公式化的系统比组织自身制订的系统更好。

建筑业无疑是庞大的和复杂的行业。在此行业当中，肯定会有一些障碍阻止质量文化的发展。竞争投标是一个最明显的例子。那些被迫降低价格以获得一个项目的承包商，将尽量使用最经济快捷的方法以求不超出预算，在这种情况下，质量明显是受害者。当分包商能够轻而易举的进入这个行业而且选择分包商时不考虑他们本身的质量时，不管总承包商拥有什么样的质量系统也都无法实施了。另外就是总承包商和分包商的管理复杂，水平不同。技能的缺乏使情况变得更糟。

不过以上这些趋向于周而复始的争论。唯一可以解决争论的办法就是文化转变，不仅要转变承包商的文化，还要转变顾客，顾问咨询公司和分包商的文化观念。作者认为，建筑业的文化转变已经开始，而且一旦获得契机，建筑业的文化改变就会比预计的还要快。

第八章

西方与中国建筑业中建设管理的当前问题

(注：本章的内容曾刊登在2001年5月25～27日在中国北京召开的项目造价管理国际研讨会论文集里)

导　言

我们认为一直有这样一种思潮，认为因为中国正处于从传统的计划经济向市场经济转变的过渡阶段，所以，中国应加速采用西方现代的管理概念，而这也许是不必要的。

也许值得强调的是，在西方，无论是广义上的管理或是专门意义上的建设管理，均没有单独统一的方法。举例来说，当英国专业人士试图使用美国的价值管理系统时，他们发现根本无法使之运作。其中原因有很多，但主要与价值工程研究的原始目标有关。美国价值管理工程系统是应政府对工程承担更大的义务和责任之需要而诞生。(在美国绝大多数价值工程行动属政府行为。)而在英国，情况则完全不同。英国工料测量系统(成本控制系统)已经提供了一切必要的责任和义务。价值工程之所以需要，是为了提供一个检验造价和是否物有所值的平台。这就是一个证明在西方国家由于文化差异而导致其对于相同的管理概念的不同的解决方法的例子。

美国的重组工程与欧洲的也存在着不同之处。欧洲人普遍认为，翰默(Hammer)和珊比(Champy)的方法过于激进，已大大超出欧洲人所能接受的范围。西门子公司(Siemens)的首脑人物，亨利其·翁·皮安(Heirich von Piener)，在《商业过程重组》中曾提及："我对于翰默(Hammer)先生的激进论文并不完全感觉舒服。我们的雇员不是中子，而是人。因此，对话至关重要。"何尔森(Holtham)亦表示，商业过程重组需要植根于欧洲的与众不同的管理特点上，以欧洲的人本主义思想为本，这与美国的更为机械化、分裂化的(管理)方法形成鲜明对比，这种方式使得企业内部、企业间、供货商与消费者之间，以及跨国界的合作化概念得以提升。何尔森(Holtham)对于欧洲商业过程重组的想象与翰默(Hammer)对于美国商业过程重组实施的描述，也许没有多少差别。不同之处是理论的表述方式而不是理论本身。正如翰默(Holtham)总结的，"商业过程重组核心要素的确定已超越了合乎新教(教义)的北美方式，如果它围绕欧洲的环境而被确立，商业过程重组将对欧洲的有价值标准"。

科文(Cowan)，现代合伙人制运动的主要建筑师之一，强调说："合伙人制不仅仅是一套目标和程序；它是一种思想状态，一种哲学。合伙人制要求合伙企业的所有干系人，既要彼此尊重、信任，合作又要各展所长"。不管事实上，合伙人制已在美国、澳大利亚以及英国获得了多么巨大的成功，它仍未被日本建筑业所认可。因为在日本，合伙人制的总体哲学思想被更多地认为是日本企业文化的一个组成部分，合伙人制经营作为一种企业过程，实际上是多余的。

一方面，我们虽然在争论着一些西方的建设管理概念也许不适用于中国的建筑业的文化，但另一方面我们却不能否认西方文化与中国文化之间存在着可以互相溶合、彼此丰富的机会。当我们初次接触中国的有形建筑市场这一概念时❶，我们被这新奇的中国这种用来解决长期存在的行政责任制问题的管理方式迷住了。有形建筑市场是一个有趣的现象。之所以如此命名，是因为它是一个透明的、可以审计的过程，一个在有形的、具体的地点，所有的官员及建筑业代表均到场参加的公开的投标过程。"有形"一词用以明确区分这一新式过程与那些关起门来秘密签订合同的"无形"市场的不同。这种在建立有形建筑市场时使用的方法，看起来与重组概念十分地吻合，正如翰默（Hammer）提倡的："随着时间的推移，公司发展出详尽的方法来处理工作。从没人曾经转回头来看看整个系统……对于重组来说，最必要的是达到总量的飞跃，而不仅仅是持续的少量的收获。"我们相信，有形建筑市场通过达到"突变性的提高"，已经运用了重组的精髓，并且是西方重组原理跨跃文化差异的一个范例。事实上采用在有形建筑市场进行招投标这一做法可以说是一个可被应用于任何市场经济的模式。

文化趋势

虽然我们已经论证过了，对于建设管理或建筑经济还没有一个世界公认的观点，但是似乎存在这样一种普遍的趋势，即在采购过程中，项目干系人的参与越来越多。（对于采购，我们是指从楼宇项目的起始阶段至楼宇可以投入使用的竣工阶段这一完整的楼宇周期）。

从英国和澳大利亚等国的市场经济体系中获得的主要信息是就是业主在激活建筑文化转变上通过采用现代管理概念，而充当的关键角色。拉谭（Latham）对此总结说："实施始于业主。业主是过程的核心，行业内其他各方必须满足其要求。"拉谭（Latham）继而建议："政府应该努力成为最佳实践业主。政府应为职员提供必要的培训来实现此目的，并做好目标管理的安排，以便为不断提高表现而施加的压力。"更为简明、直白的表达就是："业主有掌握着支票簿的权力。"强调业主更多的参与的做法，与中国建设部1993年为中国建筑业建筑市场制定的计划纲要不谋而合。

一般来说，政府业主正在考虑医甘（Egan）、拉谭（Latham）、盖尔思（Gyles）等人及中国建设部的提议，并正在通过加强对建筑业的管理来达到不断提高的目的。虽然这一主动性行动得到赞同，但对于如何实现它，还有好多模糊不清的地方。其中原因很多。其中一个于原因也许是，因为许多现行管理概念都以哲学思想为背景，要么是系统理论支持，要么是采用完整综合的方法。我们认为这共同来源使得将这些概念分门别类的难度大大提高了。

---

❶ 1993年，中国建设部正式出版了其关于建筑市场的计划纲要，此纲要以对投标的分配进行改革——控制投标过程，而非投标人为基础。继而，监察部和建设部1994年7月发出联合通知，就贪污受贿和不公平竞争问题做出部一级的联合批示。接着，国务院做出决定。该决定强调了：
- 采用建设项目注册制度；
- 引入和提倡公开招投标制度；
- 严禁政府部门在各种程度上干预招投标制度；
- 加强对投标单位的监督管理，以避免串通性的投标；
- 教育参与公开招投标过程的政府官员做到公正、诚实、可信；
- 加强对项目的监督工作，同时加强对贪污受贿或玩忽职守案件的调查和管理。
  这些部一级的行政规定所带来的一个结果就是有形建筑市场的建立。

现行概念间缺乏差异的情况在以下论述中表现得最为典型,如"可建设性不仅仅是价值工程或价值管理"或"重组正在取代全面质量管理"或更令人迷惑不解的,"合伙人制是针对项目而言,而全面质量管理是对建设公司而言的。"

现行的议题

在过去十年中,西方建筑业的文化发生了重要的变化。传统的对立的关系正发生着变化,而且主要的政府业主已认识到最佳的实践的承诺是发展良好而长久的关系的方法。

在澳大利亚州政府的有关部门,如新南威尔士州的公共工作服务厅通过既定的格标准,对承包商资格预先进行审查的做法,来主动提倡最佳实践。

公共工作服务厅对承包商资格认证列出了以下几项建筑业最佳实践标准:
1. 对业主满意程度的承诺;
2. 质量管理;
3. 职业健康及安全和再就业管理;
4. 合作性的合同;
5. 工作场所的改革;
6. 对于环境问题的管理;
7. 合伙人制;
8. 目标管理;
9. 由承包人推荐并被公共工作服务厅接受的其它最佳实践领域。

对于价值超过 2000 万澳元的合同承包商必须提供一份被认可的在承诺方面的记录,或者一份如何执行上述标准 1 到 6,以及具改革性的极佳 7 到 9 种的至少一项的公司计划书。

按照这一计划,获得最佳实践认证的承包商较之那些未获得认证的承包商,将被给予显著多的投标机会。可以预见,最终,只有那些通过此计划获得认证的承包商才会有资格参加价值超过 50 万澳元的合同的投标。(这是一个说明了市场经济中支票簿具有的威力的范例)。

列举公共工作服务厅的实例,不是为宣传这个政府部门,而是为阐明本章的中心议题,即业主正在逐步介绍现代管理概念,并正在对建筑业施加压力使这些概念得到加速的来用。他们或者直接运用一些管理概念,例如正式的重组原理,或者较为含蓄些,如在中国,采用以有形建筑市场的形式进行来彻底改革招投标方式。在 20 世纪 80 年代,建设管理从业人员,或者建设管理专业的大学生也许从没听过诸如价值管理、全面质量管理、可建设性、目标管理、合伙人制和重组等概念。到了 20 世纪 90 年代,越来越多的人认识了这些概念,而且业主,特别是有关政府部门极力主张应用这些概念。然而,关于这些概念的属性和实用性,仍存在着好多模糊不清的问题。

正如前面提到的,其原因之一也许是大多数的现代管理概念均以哲学思想体系为背景,如果不是系统理论,那么至少也依循着整体处理方法,这些方法鼓励着增加干系人的参与。

多年以来,建筑业的评论家一直关注那些已被感知的建筑业内的关于分裂和各自分隔的问题。大多数经常困扰建筑业的问题是因为没有纵览全局的能力。许多上述方法的倡导者宣称,他们的概念纠正了这一缺陷。例如,何拉德(Hellard)曾主张:"合伙人制肯定将会成为整体处理方法的关键。这套方法一定首先引入到企业当中,然后与分包商和主要承包商一起被并入到团队表现中。"对于这些观点,我们没有发现错误,大多数其它现代建设管理概念也赞同类似的观点。这些观念的融合所带来的结果就是多数建设管理概念为了引起同一决策者的注意,而相互竞争。

这一现象可以通过参考早已被广泛应用到建设管理学科的"成本影响曲线"(基于帕累托(Pareto)原理),而以图 8.1 来说明,帕累托(Pareto)原理认为个人或团体越早参与进决策过程当中,其对于项目结果的潜在影响就越大。相反的,影响项目结果的能力将随着时间以指数方式减小。我们认识到,问题在于当大量的概念为了在 x,y 轴的原点争夺到主动地位时而引发的冲突。

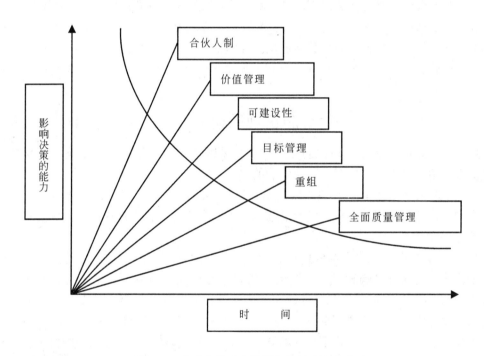

图 8.1 在成本影响曲线原点处的冲突要求

这仍是个值得讨论的问题,即虽然政府机构鼓励,并且在某些情况下,试图加强促进这些概念的应用,但对于如何同时或相结合地运用这些概念,仍缺乏适当的引导。

更让人吃惊的是,虽然大多数现代管理概念均以系统理论为基础,这却仍没能使这些概念达成一致,反而更为相悖。问题不全出在系统理论上,而是与应用这些概念的方法有关。

许多建设管理领域的研究人员都倡导系统方法。比如凯莉(Kelly)和梅尔(Male)推荐在价值管理领域中使用系统理论和系统思想;还有陈(Chen)和麦乔治(Mcgeorge)在建立可建设性模式的过程中也如此。同样可以证明的是系统方法也是重组的基础。但是,车克兰(Checkland)讽刺性的评论道:"近些年来,系统方法已成为一个时髦用语了。很少人会坦言他们在工作中不采用系统方法,而且如果一位管理学教科书的作者没有在其书的副标题位置上注明'一套系统方法',将是很不明智的。"至于系统方法究竟指什么,这一问题仍存在着很大的模糊性。系统方法经常被简单认为含有整体看法的意思。然而,车克兰(Checkland)注意到"系统范例关注全局及其特性。它是完整综合的,但与通常意义上的'整体'有所不同;系统概念关注的是整体组成及其等级安排。而不仅是整体的概念。"而我们的兴趣就集中在这些概念的等级关系这一观念上。

虽然,很明显的,诸如重组和合伙人制等概念的某些方面有着某一共同的目的,但这并不必然意味着这些概念会顺从把他们按照一些形式的等级安排进行分类排列。举个例子,重组和合伙人制都与文化改变有关,都是关于破除现有障碍,以及建立不同关系及联络方式。至少,理论上,这两个概念不是相互独立的,而且重组有可能在合伙人制的环境下发生,或者合伙人制的关系是实施重组的结果。他们的关系是复杂的,而且可能并无等级之分。既然合伙人制可在重组内部"筑巢",反之亦然,如果我们不断地往上覆加诸如可建设性和全面质量管理等概念,由于没有哪些概念相互间是必然独立的,那么随着概念连续不断地覆加,一个关于内在关系的概念性模型将会变得越来越复杂。

关于这套关系,有一个极具诱惑力的普遍性,正如我们前面谈到的,就是跨越文化的差异。

新出现的问题

帕累托(Pareto)延伸,用以说明业主在项目采购过程中日益扩大的影响,以及它是如何对西方和中国建筑业结构产生重要影响的。

图8.2过帕累托(Pareto)影响曲线显示了不同的合同关系,并表明了合同安排类型安排与采购方式随着时间的推移而发生的变化,以此来满足一直不断提高的业主要求和干系人的参与程度。

从二战结束至80年代中期的这段时间,对于建筑师而言是一段美好的时光,他们被看作是传统上的项目采购和设计团队的领头人,而承包商的参与只限于在投标之后。80年代中期和90年代早期,出现了设计连建造的集合型过程,政府业主尤其被这种设计连建造的风险共担机会所吸引。90年代宣布了建设——拥有——经营——转让项目采购形式的出现,这种形式特别强调干系人的介入和业主及承包商在项目结果中的完全参与。

图8.3通过帕累托(Pareto)曲线显示了业主的影响各个专业的上升趋势。帕累托(Pareto)影响曲线的内在逻辑就是在任何复杂的决策过程中,早期参与到过程中的参与者会拥有可以施加最大程度的影响的潜在的可能性(以及良机)。有趣的是,二战之后从专业的影响角度来说,随着设施管理专业已在90年代成为最重要的参与者,这一最大程度(的影响)带来了很大的好处。设施管理已被设施管理中心(Centre of Facility Management)定义为"一个组织为满足策略性的需求,在优质的环境中提供并保持支持性服务的过程。"换言之,是提供一个与组织的目标一致的建筑环境。全面设施管理方式对现代商务公司最为适合,那里已强烈意识到了客观的工作环境对于组织的目标的影响。从这个定义就很容易看出为何在欧洲、美国及澳大利亚,设施管理已成为建筑业中发展最快的一个专业。我们的观察是这一趋势很可能在中国出现。

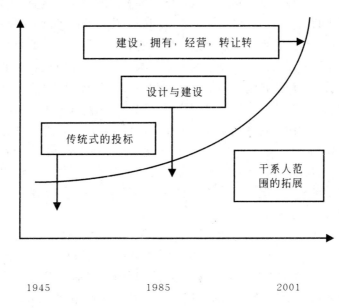

图8.2 干系人在项目采购过程中的参与程度呈指数增长

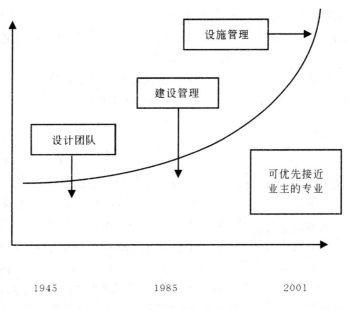

图8.3 变化中的专业的角色

未来方向

正如我们在图8.2和图8.3中试着说明的,在楼宇建造和表现过程中,业主及用户的参与活动的不断增加有可能导致用户与承建商之间的复杂关系的加速的加剧。现今,设施管理也许是一门基础的学科，并且它可以最大限度地挖掘出诸如合伙人制、可建设性、全面质量管理、目标管理、价值管理、以及重组等概念的潜能。然而，仍然不太清晰的是如何利用这些方法的综合效力，以及如何在不同的文化背景的基础上采纳、调整或者抛弃这些概念。也许对于目前建筑业中运用这些概念的方式而言，较为普遍存在的一个基本方面（或缺陷）就是业主方还未被完全整合到这个采购过程中。举例而言，虽然合伙人制有能力涵盖所有的干系人，但这种情况即使曾经发生过，也是极为罕见的。偶尔业主参与其中，但在某种程度上还是被排除在这些过程之外。比如，虽然澳大利亚新南威尔士州公共工作服务厅制订了一套专门在合伙人制过程中使用的合约(C21)，但在业主与承包商之间仍然存在着一种"我们和他们状况"的感觉。甚至在建造——拥有经营——转让形式的采购过程中，业主（通常指政府当局）仍存在着某种程度的分离感，虽然只在经营阶段之后才占有该设。这意味着，在某些情况下，这类项目的经营期会延长达 40 到 50 年间（正如最近竣工的悉尼东区收费高速公路这一例子）。

在业主与发展商全面参与分担风险方面，一个有趣的发展（方向）就是联盟日益得到普及。杜子(Doz)和翰摩尔(Hammel)曾将联盟的发展描述为现代公司发展当中最为活跃的特征"几乎每天都会有新的联盟诞生。"这些联盟被证明是灵活性组织机构的流行趋势，并且是对变化中的公司环境的最直接的反应。据经济学家情报组(Economicist Intelligence Unit-EIU)称，各类型联盟到 2010 年时将成为最重要的管理工具之一。经济学家情报组(EIU)的调查显示，29%的公司有介入联盟，而且根据预测，这个数字到 2010 年时会迅速激增至不少于 63%。

联盟过程被葛利伯(Gerybadze)描述为"……业主和与之相关的公司将为一特定具体项目而共同努力，但（各自）保留合法的独立的机构。进行合作的公司的所有权及管理（方式）将不实行完全整合，虽然项目的风险由所有的合作者共同分担。"

也许对项目联盟安排下的资金计划安排的描述就是对联盟实质的最佳阐述。联盟关系的一个具体特性盈利均摊/亏损均摊的方法。如图 8.4 所示。

盈利均摊/亏损均摊

根据这个联盟协议，盈利分摊/亏损分摊计划将风险、毛利以及合作各方的公司相关的支出按照联盟的总体表现和最后成本进行分配。如果成本出现超额支出，参与者要拿出利润及企业开支总和的 50%直至最大值；但如果存在成本结余，那么这些成本结余的 50%将被还给参与者，没有上限，因为不会有人期望减少奖励。没有其它处罚性规定被包含在联盟协议中，因为业主知道分摊利润和相关成本，以及分摊患难亏损足够的激励因素了。参与公司之间分摊盈利或亏损的原则是 50%归于业主，而另外的 50%按比例和目标成本分配给其他参与公司。如果竣工时的项目实际成本比目标成本低，则此额外盈利将按照各参与者在其盈利分摊/亏损分摊协议中所占百分比分配给他们。如果项目支出超出目标成本，则所有的参与者，包括业主，均要按其在盈利/亏损均摊协议中所占的百分比对超支负责。

我们认为，联盟的利润刺激是市场作用文化与合伙人制文化特质之间的一种有趣的结合，并且我们相信，这对中国建筑业颇有借鉴意义，因为中国建筑业正试图通过市场力量改善建设项目的时间、成本及质量，并为干系人创造最佳效益。

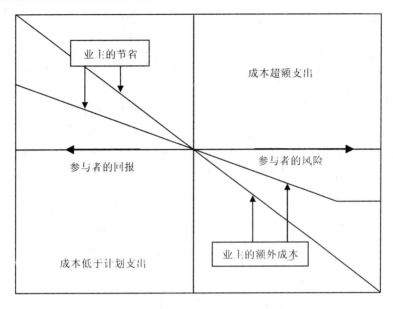

图 8.4　盈利/亏损均摊的联盟协议

结　论

利格比(Rigby)在其《过程重组的秘史》中假定了这样一个概念：所有的管理概念都会经历六个阶段，即以前概念的不足；解决方法的发现或再发现；早期成功事迹公开而产生的愉悦感；由于在不适当的情况下过分使用该技术所导致的超范围(运作)；失败的例子逐步增多，以至不能忽视的程度时所引发的嘲笑；因为该技术被抛弃或被新技术所代替而最终放弃该技术。当我们试图对目前及未来建设管理和经济问题进行阐释时，我们尽量回避这类型的挖苦话，但我们意识到也许我们揭示出的问题远比提供的解决方法多。在过去的几十年间，企业组织结构的改变是显而易见的(不论是在西方国家或是中国)，同样，那些经理们在领导和引导这些进化中的企业组织时所应用的方法也是不断改变的。在本章开篇处，我们曾描述有形建筑市场的建立，是对于中国建筑业中的困难及已确定的问题的有趣并富于创新的反应。我们还对西方管理方法中出现的文化转变和在使目前相互竞争的管理概念和谐一致时遇到的一些困难进行了讨论。

我们通过以下建议做为总结，即那些全球性的不可预知的社会性的、经济技术上的以及政治方面的因素，将迫使企业去寻求建立国家的、乃至跨国的商务联盟，并且，我们已提及了联盟的出现是其中的一条出路。当然，通过我们对于西方和中国建筑业环境的了解，我们相信双方都会从对方那里获益匪浅。

Construction Management in a Market Economy

Denny McGeorge
Angela Palmer
Patrick X.W. Zou

Acknowledgements

The nature of this book has meant that we have drawn on a large number of other resources and we acknowledge our indebtedness to all of them for their kindest support by allowing us to use their valuable resources, in particular, Takayaki Minato, John Kelly, Professor Vernon Ireland, Dr Selwyn Tucker and his colleagues, Professor Tony Sidwell and his colleagues, Glen Peters, Professor Chen Swee Eng. We also acknowledge our indebtedness to many commentators in construction management and management science who are too numerous to mention by name, but without whom this book could not have been written.

The authors would like to thank the Faculty of the Built Environment, The University of New South Wales for providing funding for the book writing.

Preface

I am delighted that Professor Denny McGeorge and Dr Patrick Zou of the University's Faculty of the Built Environment together with Dr Angela Palmer have as joint authors published this new book on Construction Management in both English and Chinese. All three authors have considerable experience working in Asia.

The University has a proud record of collaboration with its many friends and colleagues in China over a vast range of professional activities in both academic and governmental links and commercial enterprises. This growing partnership of co-operation builds on the expertise and experience that China and Australia possess, enabling special skills to be shared between the two countries.

The book is designed to assist and help students as well as practitioners in the vital area of the built environment and it will assist in the growing interchange of ideas and research linkages between China and Australia.

The University congratulates the authors and the Faculty on the signal achievement of publishing this book and of publishing it in Chinese and English.

John Yu *AC*
Chancellor
The University of New South Wales
Sydney Australia

Foreword

The purpose of writing this book *Construction Management in a Market Economy* is to introduce readers, in China, to the Western approach to procuring buildings. By procurement we mean the complete building cycle from the inception stage of a building project to the completion stage when the building is ready for occupation.

The three authors of this text, Emeritus Professor Denny McGeorge, Dr Angela Palmer and Dr Patrick Xiaowei Zou have a collective experience of both Western and Chinese approaches to construction management. Hopefully this experience has made us sympathetic to the needs of our Chinese readership. This book is not however an instruction manual of how to procure buildings in China. It would be presumptuous of us to make this claim. What the book does attempt to do, is to explain construction management techniques, as currently practised in the West, in a way that is meaningful to Chinese readers.

The book is intended to be topical and it brings together in a single volume the main new management techniques: *Benchmarking, Reengineering, Partnering, Value management Constructability* and *Total quality management.* In the final chapter, "*Current construction management issues in Western and Chinese construction industry*", we have discussed some issues which are of current concern to the Chinese construction industry.

The book provides an objective account of management concepts and demonstrations how they interrelate. It will be of interest both to postgraduates in construction management and final year undergraduates in construction management and surveying, and to practitioners needing a readable introduction to these concepts.

It is perhaps worthwhile emphasising at this point, that there is no single unified Western approach to construction management. For example, when British companies tried to use the US system of value engineering they found that they couldn't make it work. The reasons for this were numerous but largely related to the original objectives of the value engineering studies. The US system of value engineering was born out of a need for greater accountability on government projects. Almost all value engineering activity in the US is government work. The situation in the UK was very different. The UK quantity surveying system (cost control system) provided all the accountability that was needed. Value engineering was required to provide a platform for the examination of value as opposed to cost. This is an illustration of how cultural differences between Western countries has lead to different approaches to the same management concept. Differences also exist between the American approach to Reengineering and the European approach.

It is to be anticipated that China will also develop its own versions of construction management concepts such as Reengineering, Total Quality Management, Benchmarking and Value Management. It may well be the case that some Western concepts may prove to be inappropriate in China due to cultural differences. We have not however filtered out Western concepts or management techniques on the grounds that some of these may not be suitable for the Chinese construction industry. Given that the Chinese construction sector is in a transitional phase of moving from the planned to the market economy we believe that it is important, in the first instance, that all concepts are considered even if in the longer term some of these concepts are not adopted.

It is also worthwhile emphasising at this point that the Chinese construction industry is also developing its own unique form of procurement with the introduction of the Tangible Construction Market (TCM). The TCM is an excellent example of process re-engineering and is discussed in detail later in this text.

In order to facilitate the Chinese readership, we have translated the text into Chinese and included the Chinese translation in this book. The fact that China has recently joined WTO (World Trade Organisation), means it is necessary to refer to the Western management techniques while enhancing their own strength to be more sustainably competitive. We sincerely hope that our efforts in producing this book will be of use to the readers, either in their research or construction project management practice.

TABLE OF CONTENTS

Preface

Foreword

1 Introduction 1
    The book's contens 2

2 Benchmarking 7
    Introduction 7
    Definition of benchmarking 9
    Historical development 10
    Types of benchmarking 11
    The process of benchmarking 14
    The benchmarking team 28
    Benchmarking Code of Conduct 29
    Legal considerations 30
    Benchmarking: the major issues 30
    Current research 32
    A case study 34
    Conclusion 34

3 Reengineering 39
    Introduction 39
    Origins of reengineering 40
    Reengineering in a construction industry context 41
    The goals of reengineering 42
    Reengineering methodology 45
    Pitfalls of reengineering 51
    Information technology and reengineering 54
    Reengineering from a European perspective 57

A case study of a process reengineering study in the Australian construction industry 58
Conclusion 68

4 Partnering 71

Introduction 71
The origins of partnering 71
Partnering in a construction industry context 72
The goals of partnering 73
Categories of partnering 74
Project partnering 74
Partnering Charter 81
Strategic or multi-project partnering 86
Legal and contractual implications of partnering 89
Dispute resolution 92
Conclusion 92

5 Value management 97

Introduction 97
Historical development 97
Function analysis 102
Organisation of the study 110
Who should carry out the study? 111
Who should constitute the team? 111
The format of the study 113
Where should the study be carried out? 114
The timing of the study 114
How should alternatives be evaluated 115
Value management as a system 117
The American system 117
A case study of value management in the United States 118
The British system 120
A case study of value management in the UK 120

	The Japanese system 124
	A case study of value management in Japan 124
	Why are the systems different? 126
	The relationship between value management and quantity surveying 128
	Conclusion 128
6	**Constructability 133**
	Introduction 133
	Origins 133
	The goals of constructability 134
	Implementing constructability 137
	Constructability in practice 140
	Constructability and the building product 142
	Good and bad constructability 146
	Quantifying the benefits of constructability 151
	Conclusion 152
7	**Total quality management 155**
	Introduction 155
	Definition of TQM 157
	What is quality? 158
	Historical development of TQM 161
	The need for a paradigm shift 163
	A change in the culture of the construction industry 165
	Customer focus 167
	Integration 168
	The all-embracing nature of TQM 174
	Continuous improvement 174
	Quality costs and the cost of quality 175
	Universal standards of quality such as ISO 9000 176
	Change management 177
	The methods of TQM 177
	How to implement TQM 178

Kaizen 179
Current research into TQM in the construction industry 179
Conclusion 180

8 Current construction management issues in Western and Chinese construction industries 183

Introduction 183
Cultural trends 185
Current issues 186
Emerging issues 189
Future directions 191
Conclusion 193

Bibliography 195

Benchmarking 195
Reengineering 195
Partnering 197
Value Management 198
Constructability 198
Total Quality Management 199

1   Introduction

The construction industry now faces substantial demands for improvements in quality and cost control and a reduction in contract disputes. A host of new management techniques has been promoted to help achieve this, but many in the industry find the concepts confusing and are sceptical about their usefulness.

Our purpose in writing this book was twofold. Firstly, we felt that there was a need to bring together, for the first time in a single volume, most, if not all of the management concepts currently being advocated for use in the construction industry. The concepts which we have selected for detailed scrutiny are: *benchmarking; reengineering; partnering; value management; constructability* and *total quality management*. Although we do not claim that this is a completely exhaustive coverage of the field, it is extensive and is more than sufficient for the second purpose of this book, which is to address the challenge of how to achieve a synergistic result from the multiple application of management concepts.

We are aware, from conversations with industry practitioners, that a good deal of healthy scepticism abounds with respect to the efficacy of modern management concepts. Many practitioners would be sympathetic to the view[1] that management concepts pass through a sequence of six phases. These are: deficiency of previous concepts; discovery or re-discovery of a solution; euphoria as early success stories are publicised; over extension due to the excessive application of the technique to inappropriate situations; derision as examples of failure grow too large to ignore; final abandonment as the technique is discarded or replaced with a new technique. Whilst we do not subscribe to this viewpoint, we feel that it is important to distinguish genuine cultural shifts in the industry from what can simply be trendy ideas. We are not alone in making this distinction, Godfrey[2] was also alert to this issue arguing that 'The use of partnering is growing fast, but there is a danger that this will be merely a passing fad'.

The book has been written against the backdrop of an industry who's decision takers now have an arsenal of management concepts at their disposal. Whilst the worth of such concepts such as total quality management, value management and benchmarking, are not in dispute, there is a real danger that practitioners in the industry are becoming saturated and disillusioned by the number of management concepts which have emerged in the last decade, and with the apparent overlap which exists between them. In a recent survey carried out in Hong Kong[3], it was found that a significant number of construction industry clients thought that 'value engineering was another name for a cost saving exercise, buildability study or one of the techniques of cost control'. This situation has been exacerbated by the fact that each new management concept has its own set of protagonists who advocate the adoption of 'their' concept to the exclusion of all others. Thus, from a practitioner's perspective, some concepts may appear to be mutually exclusive rather than complementary, with each new concept jockeying and jostling for the attention of the key decision takers. This aggressive promotion of a concept is typified by Hamer's slogan[4] for re-engineering 'don't automate, obliterate' or Kelada's rhetorical question[5] 'is re-engineering replacing total quality?'

We have attempted in this book to give a straightforward and objective account of the chosen concepts. This book is to inform industry practitioners and academics of the state of the art of construction management concepts and at the same time to provide a conceptual model which makes for a better understanding of the inter-relationship of current concepts and of concepts yet to be developed.

Throughout the book we have diligently tried to differentiate between concept as defined in *Webster's dictionary*[6] as an abstract idea generalised from particular instances and technique, defined as a method of accomplishing a desired aim. The book deals in detail with both concepts and techniques, the philosophical learning is however, towards the conceptual.

The book's contents

Chapter Two: Benchmarking

Benchmarking is a concept aiming at improving the competitiveness of organisations through the examination and refinement of its business processes. The concept has in origins in Xerox Corporation who stripped down copiers manufactured by its competitors and compared them to their own. They later extended this comparison to include business processes of its competitors. The chapter looks at *types of benchmarking; the process of benchmarking; the benchmarking team and the benchmarking code of conduct.* The chapter concludes by illustrating a simple case study of benchmarking the customer focus of a national house-builder against a national car manufacturer.

Chapter Three: Reengineering

Reengineering is being hailed as a management revolution, which could have repercussions on the scale of the industrial revolution which followed Adam Smith's *Wealth of Nations*. The proponents of business process reengineering are claiming quite dramatic results following its introduction. The full impact of reengineering is yet to be felt in the construction industry, although interest is gaining ground. The following aspects of reengineering are covered: *origins; reengineering in a construction industry context; goals; methodology; implementation; time and cost saving; pitfalls; IT and reengineering; a European perspective; and the T40 project.* The chapter contains a detailed case study, known as the T40 project, of the initiation, planning and implementation of a process reengineering in the Australian construction industry. The objective of the project was the reduction of construction process time by 40%.

Chapter Four: Partnering

The concept of formal partnering is of relatively recent origin, dating back to the mid 1980's. The concept was developed in the United States and has spread to other countries including Australia and New Zealand in the Southern Hemisphere and also to the UK. Parties adopting partnering resolve to move away from the traditional adversarial relationships to a 'win-win' situation. Partnering can either be undertaken at the level of a single project and be relatively short duration or can be of a semi-permanent nature at a strategic level. The chapter traces *the origins of partnering; partnering in a construction industry context; the goals of partnering; categories of partnering, project and strategic; the participants; commitment; the partnering process; how to conduct partnering workshops; partnering charters; the pitfalls of partnering; limits to partnering; legal and contractual implications of partnering; and dispute resolution.* The chapter ends by speculating on the likely uptake of partnering by the construction industry as a whole.

Chapter Five: Value management

Value management was developed in the United States manufacturing industry during the Second World War. Its aim was to improve the value of goods by concentrating on the functions that products perform. It was so successful in manufacturing that the United States Department of Defence began using it in the construction industry and it was around this time that an interest in value management was shown by the British construction industry. The chapter traces *the historical development of value management; the use of function analysis; organisation of value management studies; the evaluation of value management proposals; the American system of value management; the British system of value management; and the Japanese system of value management.* The chapter ends by analysing why these three systems are different and examines some of the major cultural influences on value management development.

Chapter Six: Constructability

Constructability is the only concept in this book which is the exclusive domain of the construction industry. Constructability is concerned with how decisions taken during the procurement process facilitate the ease of construction and quality of the completed project. From its inception in the early 1980's constructability has moved from its original narrow focus to incorporate decision support theory and decision support systems. The following aspects of constructability are covered: *the origins; scope and goals; implementation; constructability in practice; the building -in use; good and bad constructability - indicators of success; and quantifying the benefits of constructability.* The chapter concludes by distinguishing between constructability and good multi-disciplinary team working.

Chapter Seven: Total quality management

Total quality management or TQM is a concept aimed at improvement of the organisation through increased customer focus, integration of the organisations processes and a philosophy of continuous improvement. The chapter examines *definitions of TQM; historical development; the need for a cultural change in the construction industry; customer focus; integration; continuous improvement; quality costs and quality standards.* Finally the chapter briefly examined the array of quality methods that are currently available.

Chapter Eight: Current construction management issues in Western and Chinese construction industries

The chapter takes as its theme, the dominant message from market economies such as the United Kingdom and Australia that the key role of the government client is in activating a cultural shift in the industry through the strengthening and improvement of management practices. In Western economies this has lead to the adoption of management techniques such as benchmarking, total quality management, constructability, value management, partnering and reengineering. The counterpoint in China has been the introduction of initiatives such as the Tangible Construction Market (TCM). In this chapter we present a view of the current situation in which the burgeoning number of Western management concepts is seen to be a potential problem. The growth of facilities management and the changing relationships of the design and construction professions with the client are charted using the Pareto influence curve. We suggest that unpredictable social, economic, technical and political aspects of a globalising society will force organisations to look at forming national and trans-national business alliances. The proposition is made that project alliancing is one management practice which will become increasing popular in the coming decade. The chapter concludes by expressing the view that increased dialogue between Western and Chinese construction industries is essential for the common good and, that both systems have much to gain and learn from each other.

As mentioned in Foreword, it is perhaps worthwhile emphasising at this point, that there is no single unified Western approach to construction management. For example, when British companies tried to use the US system of value engineering they found that they couldn't make it work. The reasons for this were numerous but largely related to the original objectives of the value engineering studies. The US system of value engineering was born out of a need for greater accountability on government projects. Almost all value engineering activity in the US is government work. The situation in the UK was very different. The UK quantity surveying system (cost control system) provided all the accountability that was needed. Value engineering was required to provide a platform for the examination of value as opposed to cost. This is an illustration of how cultural differences between Western countries has lead to different approaches to the same management concept. Differences also exist between the American approach to Reengineering and the European approach.

It is to be anticipated that China will also develop its own versions of construction management concepts such as Re-engineering, Total Quality Management; Benchmarking and Value Management. It may well be the case that some Western concepts may prove to be inappropriate in China due to cultural differences. We have not however filtered out Western concepts or management techniques on the grounds that some of these may not be suitable for the Chinese construction industry. Given that the Chinese construction sector is in a transitional phase of moving from the planned to the market economy we believe that it is important, in the first instance, that all concepts are considered even if in the longer term some of these concepts are not adopted.

---

[1] Rigby D. (1993) The secret history of process engineering, *Planning Review* March/April; 24-27.

[2] Godfrey K.A. Jr., (editor). (1996) *Partnering in design and construction.* New York: McGraw-Hill.

[3] Fong P.S.W. (1996) VE in construction: a survey of clients' attitudes in Hong Kong. *Proceedings of the Society of American Value Engineers International Conference*; Vol. 31.0.

[4] Hammer M. (1990) Re-engineering work: Don't automate, obliterate. *Harvard Business Review,* July/August; 104-112.

[5] Kelada J.N. (1994) Is re-engineering replacing total quality? *Quality in progress,* Dec; 79-83.

[6] Webster's Third New International Dictionary 1976 Edition, Encyclopedia Britannica, Chicago.

2 Benchmarking

Introduction

In 1996 Michelle Smith, an Irish swimmer, confounded the swimming world by winning three gold medals at the Atlanta Olympic Games. When asked about the secret of her success she said, among other things, that she had learned the training methods of track and field athletes and applied these to her swimming training programme. Michelle's experience is not new. Other athletes have also looked outside their own sports for new techniques that have formed the basis of very successful training programmes. Emil Zatopeck, who was the only man ever to win three long distance athletic gold medals at one Olympic games, learned his techniques from the Army[1]. Other athletes such as Ron Hill, one of the world's greatest marathon runners, used the carbohydrate loading diet invented by Swedish physiologists to improve his performance[2]. Others are said to have lived on a diet that included turtle blood and ground rhinoceros horn.

What is common in all these experiences is that individuals looked outside the scope of their own sports or disciplines to find ways of improving. They were using the training methods already accepted in their own sports but these were not enough. Everybody was using them. In order to really succeed they needed something else. They needed a competitive edge.

Parallels for this can be seen in industry. For example when Henry Ford II was faced with rescuing a failing business he took new concepts of management from his competitor, General Motors[3]. However although there are examples like Ford, industries and companies are reluctant to look beyond their own sphere in order to find the competitive edge. For reasons of competitive fear, lack of resources or simple conservatism organisations tend to rely on the tried and trusted methods that exist within their own limited spheres. This is not to say that these tried and trusted methods are worthless: Michelle Smith was an international swimmer even without the track and field methods she introduced into her programme.

The message therefore is that the search for superiority is a three layered pyramid of success. In the case of the athlete it means they first must do the best that they can. Second they must do the best that others in their field can, by studying the training methods of other athletes. Finally they must do the best there is by looking to the outside world and examining techniques in the fields of physiology, psychology and nutrition and apply these to their own training programmes.

This idea also applies to the management of a company. In order to gain competitive edge a company needs to look at itself first. It needs to examine its own systems and methods of working and make necessary improvements. It also needs to look at its own industry to learn the best methods from it and try to achieve those best practices itself. Finally it needs to look outside its own industry to learn the best methods from other industries and to try to achieve those best practices also.

This process of looking outside one's own sphere, be it to other divisions, companies or industries, is basically one of comparison. It would be pointless investigating other companies if the information gathered were not used as a standard against which to measure ones own performance. Once this comparison is made, however, and a performance gap established, it can be used as a basis for setting goals aimed at the of improvement of one's own practices.

This pyramid of success based on the comparison with others is the basis of benchmarking (Figure 2.1). Benchmarking is the comparison of practices either between different departments within the company, or with other companies in the same industry, or finally with other industries. The aim of benchmarking is to achieve superiority.

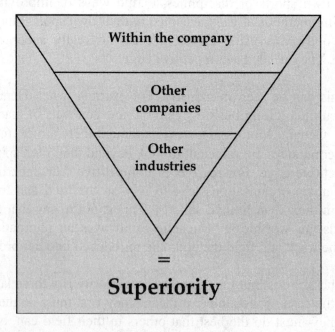

Figure 2.1 Benchmarking: The pyramid of success

Definition of benchmarking

What makes benchmarking different from other management techniques is the element of comparison, particularly with the external environment. However benchmarking is more than simple comparison and before defining it formally it is worth considering its other essential ingredients.

It is structured

Some managers will instinctively compare their departments or companies with others. Comparison may also happen inadvertently when, for example, trade journals publish league tables or statistics. This however is not benchmarking. Benchmarking is a management technique aimed at achieving superiority. As such it must be a formal and structured approach that is planned, implemented and monitored.

It is ongoing

The business environment within which a company operates may change quickly. New products, processes and techniques are constantly emerging and in order to maintain a position of superiority companies must respond to these changes. For this reason benchmarking, in order to be effective in reaching its aim of superiority, must be a continuous process.

The comparison is with best practice

If a company simply compares itself with others it will not necessarily improve. A critical factor in benchmarking therefore is the search for best practice with which to compare. It is only by benchmarking against best practice that superiority can be achieved.

Its aim is organisational improvement through the establishment of achievable goals

Benchmarking against best practice illustrates to a company the difference between its own practice and best practice. As such it allows the gap in performance to be gauged and for targets to be set aimed at closing the gap and eventually achieving superiority.

Taking these essential ingredients onto account benchmarking can therefore be defined as:

> A process of continuous improvement based on the comparison of an organisations processes or products with those identified as best practice. The best practice comparison is used as a means of establishing achievable goals aimed at obtaining organisational superiority.

Historical development

The meanings of the words 'benchmark' or 'benchmarking' are known to most people in the construction industry as a level against which other heights are measured. However in management the word has been used in a broader context of setting a standard against which to compare for a considerable period of time. Even the work of Taylor who, as far back as the 1800s, encouraged the comparison of work processes, has been compared with benchmarking[4]. The term benchmarking is also used extensively in the computer industry to illustrate a level of performance of software or hardware[5].

The word benchmarking as defined above and used in the context of this book is a new management technique 'invented' by the Xerox Corporation in 1979[6]. Xerox were aware that the photo-copiers were being sold in Japan for the same price that they could be made in America and they wanted to find out why. They therefore bought the rival copiers, stripped them down and analysed their component parts. This process proved a successful means of improvement and Xerox therefore extended the use of the benchmarking technique to all business units and cost centres of the company. The non-manufacturing departments within Xerox initially found difficulty in applying the benchmarking technique, since as they only dealt in business processes they had no product to strip down and compare. However they eventually recognised that processes were the means by which the final product was delivered and that these processes could equally be compared with the external environment as a means of bringing about improvement.

Despite its success at Xerox, benchmarking remained outside of the public domain for some years. Two events helped to change this. First was the book written by Robert Camp in 1989. Camp had worked with the benchmarking initiative at Xerox for seven years and in 1989 formalised his ideas in *Benchmarking: The Search for industry best practice that leads to superior performance* . Second was in 1992 when the Malcolm Baldrige National Quality Award, which is a prestigious quality award given to American companies introduced a category of benchmarking and competitive comparisons as a criteria of the award. These two events therefore brought the subject of benchmarking into the public domain in the United States of America[7].

The increase in importance of benchmarking in the United States was followed by an increased interest in the UK and in Europe. In the UK for example the Department of Trade and Industry developed a business-to-business exchange programme offering visits to UK exemplars of best practice in manufacturing and service industries. For a nominal fee companies may visit the host organisation with the objective of transferring best practice to their own organisation[8]. The European Foundation for Quality Management now also recognises the importance of benchmarking[5].

As is often the case with the introduction of new management techniques the construction industry lags behind that of manufacturing. However benchmarking is now beginning to be researched and used within the construction industry as well. The best examples to date are the Construct IT Study[9] and the work by the Building Research Establishment (BRE)[10]. The former involved benchmarking 11 leading construction companies' use of IT in site processes against a company identified as best-in-class (best practice). The BRE work is concentrating on producing a benchmarking methodology for the construction industry.

Types of benchmarking

Earlier in this chapter (Figure 2.1) the benchmarking pyramid of success outlined that an organisation in the search for best practice against which to compare could look to internal sections or divisions, other companies, or other industries. Naturally these three types of comparison would involve different procedures and would, in addition, offer different benefits and disadvantages. For this reason they are generally classified as three distinct types of benchmarking each of which is examined separately below. Before this examination, however, another concept, central to the idea of benchmarking, must first be discussed. This is the concept of the business process.

Any organisation is broken down into a series of functions. In business terms function refers to the performance of a particular section of the organisation such as marketing, estimating or buying. All functions have an output or deliverable. In the case of the estimating function for example, the output maybe the total number of submitted bids. A business process on the other hand refers to the action that takes place within the function. In the case of estimating, the business processes may therefore be the decision to tender, the obtaining of sub-contractor quotes or the final submission of bid. Within these processes there will also be sub-processes. In the case of final submissions of bid these sub-processes might include checking of subcontractor bids, calculation of attendance on subcontractors, addition of contingency, addition of overheads and profit and submission to the client. A process differs from a function in that it is a state of being in progress or that which converts input to output. The sum total of the outputs of all the processes is the product or deliverable of the function. The sum total of all products delivered by all functions is the final product, which in the case of construction is the completed building.

In the example above the output of the estimating function can be viewed quantitatively in the amount of projects won as a percentage of those bid for. This could then be compared with other contractors to see if there was a performance gap. This type of quantitative analysis in benchmarking is called a metric. The problem is that even when a metric indicates that a performance gap exists it gives no indication of why. If on the other hand the business processes were analysed then the reasons for the performance gap would be clear from the outset. Most benchmarking texts therefore recommend that processes be examined in preference to metrics.

This however leads to a tautological problem of how a company can know there is a performance gap unless they examine the metrics first. This is where benchmarking requires a shift in the usual mode of thinking. For reasons that are beyond the scope of this text, metrics and concentration on them as a means of improvement, can tend to misdirect effort. If on the other hand the business processes and sub-processes are viewed as pieces of a jigsaw, with the picture being the function products and ultimately the final product, then it can be seen that an improvement in all the processes, or at least those most critical to the success of the organisation, will lead to an improvement in the final product. Processes give products. It can therefore be assumed that best possible processes will lead to best possible final products.

In manufacturing some benchmarking may still be based on the comparison of final products and their components. However in construction management and in this text, the word benchmarking is assumed to relate to the process only. In this context there are three types of benchmarking.

Internal benchmarking

This is the comparison of different processes within the same organisation.

As outlined earlier an essential component of benchmarking is the search for best practice in the external environment. If this is so, then why would an organisation carry out internal benchmarking? The answer to this is fourfold. First, it is possible that within the same organisation business processes will vary. This may be for reasons of location or may be historical, the company having been subject to take-over bids or mergers. Internal benchmarking gives the organisation an understanding of its own performance level. It allows best practice that exists within the organisation to be identified and installed company wide. The second reason is that internal benchmarking provides the data that will be required at the 'external' benchmarking stage. Third, internal benchmarking, by encouraging information exchange and a new way of thinking, ensures that the process of benchmarking is understood by those who will be involved in later 'external' benchmarking exercises. Finally, benchmarking is based on the examination of the business process. As such comparison with other divisions of the same company, although internal to the organisation, may provide a comparison that is external to the process under consideration.

An example of internal benchmarking would be a construction company comprising a major works division, a housing division and a refurbishment division, comparing the way the three divisions deal with the hiring of plant.

Competitive benchmarking

This is a comparison between the processes of companies operating within the same industry. The big advantage with this type of benchmarking is applicability. It is highly relevant to compare the marketing operations of two companies offering the same product and working within the same client base. The problem however with competitive benchmarking is that because we are dealing in process, the best practice of a competitor is not necessarily good enough. A particular construction company may, for example, have an excellent reputation for design and build projects. There is however no direct follow-on from this that their estimating processes are any better than others. As a result benchmarking such processes will not create superiority. In order to identify best practice in the business process, it is sometimes necessary to go beyond the sphere of one's own industry.

Generic benchmarking

This compares the business processes of organisations regardless of the industry they belong to. Some business processes are common to all industries, purchasing and recruitment are two examples. The advantages of generic benchmarking are that it breaks down the barriers of thinking and offers a great opportunity for innovation. It also broadens the knowledge base and offers creative and stimulating ideas. The disadvantages are that it can be difficult, time consuming and expensive.

Before moving on to examine how the technique of benchmarking operates, two other items need to be considered. The first of these is whether internal benchmarking is prerequisite to competitive benchmarking and whether that in turn is prerequisite to generic benchmarking. The basic answer to this is no. The benchmarking pyramid of success shown earlier is an ideal situation; all organisations should fully understand both their own processes and those of their industry before they begin to examine those of other industries. However it is possible, although not recommended, that a company could carry out a generic benchmarking exercise without having carried out either internal or competitive benchmarking. In addition, a company may carry out only internal benchmarking without recourse to the other two types.

The second item which needs to be mentioned is that not all current benchmarking texts use the same terminology and this may cause a certain amount of confusion. Table 2.1 summarises the different terminology used in some of the major benchmarking texts.

Table 2.1 Benchmarking terminology

Author	Within the organisation	Product-to-product comparison	Different companies in the same industry	Different industries
Camp[6]	Internal	Competitive	Functional	Generic
Spendolini[7]	Internal		Competitive	Functional (Generic)
Karlof & Ostblam[11]	Internal		External	Functional
Blendell et al[12]	Internal		Competitor and functional	Generic
Codling[5]	Internal		External or best practice	External or best practice
Watson*[4]		Reverse engineering	Competitive	Process
Peters[13]	Internal	Benchmarking	Benchmarking	Benchmarking
This text	Internal	Not applicable	Competitive	Generic

(* Watson also includes two other categories of strategic benchmarking and global benchmarking.)

The fact that different terminology is used to describe what is essentially the same does not matter. The important item is that an organisation selects the terminology it is most comfortable with and uses this consistently. The choice may or may not correspond with the terminology of this text summarised in Table 2.1.

Figure 2.2 shows that as the type of benchmarking moves from internal to generic the level of difficulty, the time taken and cost incurred increase along with the creativity and the opportunity for improvement. Conversely when moving from the generic to internal benchmarking, cost, time and difficulty decrease as do relevance, ease of data collection, applicability and transferability of results.

The process of benchmarking

Although the methods suggested by the major texts for implementing benchmarking studies are numerous, all the methods contain the same essential ingredients. This book suggests a nine-step approach to benchmarking as shown in Figure 2.3 and each step of this approach is examined separately.

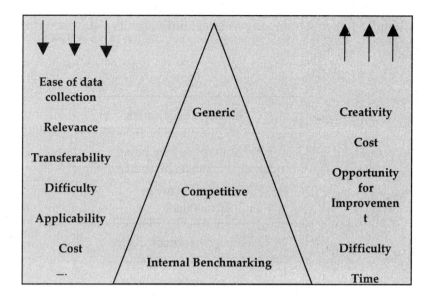

Figure 2.2 Types of benchmarking

Step 1: Decide to benchmarking

The free market is dependent on the customer being able to choose between alternative suppliers[9]. This means that when an organisation supplies goods and services that are not wanted, or not of an acceptable standard, customers will not buy them and the company will go out of business. This same freedom of choice does not exist within an organisation. The internal nature of the environment means that where goods, or more likely services, are offered by one department to another, there is no alternative available if that service proves to be unsatisfactory. What this in effect means is that although a company may appear profitable and geared to user needs, there may in fact be scope for improvement within the internal processes of the organisation. Improving these internal processes will increase efficiency and ultimately improve the standard of the final product. Benchmarking therefore provides a 'safety net' for those processes which are not exposed to market forces. As illustrated above efficiency decreases with distance from the final product[12], and this is the major reason for carrying out benchmarking. However there are other reasons, some of which are outlined below.

Speed of change

Today's business environment changes more rapidly than ever before. In the construction industry new procurement methods, new products and new clients are constantly coming to the fore. All of these exist in the external environment. A company that does not look to the outside for these will be overtaken by the pace of change, stagnant and eventually go out of business. One of the essential ingredients of benchmarking is comparison with the external environment and as such the technique forces a company to constantly appraise the changing situation of the industry.

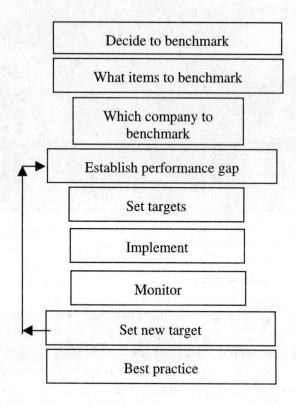

Figure 2.3 Nine-step benchmarking approach

Free exchange of information

In Britain business culture is such that information is not exchanged freely and there tends to be an underlying assumption that most information is confidential. This is totally different from the business culture that exists for example in the United States of America, where information is freely exchanged. The onset of the 'information age' will undoubtedly require a shift in the way information is regarded in Britain and use of benchmarking encourages a freer exchange of information.

External focus and meeting of customer objectives

In order to stay in business the main objective of any company must be to meet customer requirements. However, as with change, customer needs and markets are external to the company. An internally focused company cannot understand its customer needs and cannot therefore meet the demands of its markets. Benchmarking encourages the external focus required to meet customer objectives.

Recognised importance of the process

After the Second World War the need to provide high quantities of consumer goods was the main objective of manufacturing companies. Even in the construction industry emphasis was on the provision of large amounts of accommodation, both for residence and industry. This situation has changed gradually over the last 30 years. Japanese industry was quick to recognise this change and began producing goods of a much higher quality than its competitors, resulting in a huge expansion of Japans manufacturing base. Examination of Japanese methods of manufacture showed a much greater emphasis on the processes, which ultimately deliver the products. British industry is now imitating this with initiatives like total quality management, reengineering and benchmarking, all of which recognise the importance of improving the final product through examination of the process.

Highlighting the performance gap illustrates the need for change

Traditional methods of business evaluation tend to compare the performance of individual departments within the same company. Alternatively performance evaluation is based on a comparison with previous years or against a forecast of predicted performance. This type of evaluation is internally focused. Since the measure being used is either the best the company has at present, or the best they have achieved in the past, such an approach may encourage companies to fall short of the performance they are in fact capable of achieving. The true measure of a company's performance can really only be gauged by comparison with its competitors. Benchmarking, by operating in the external environment, illustrates the true performance gap. In doing so it highlights the need for, and the motivation for, change.

Best practice reveals how change can be achieved

In benchmarking the term metric is used to describe the quantifiable output of a process. For example, two construction companies will have different processes for the calculation of a tender sum based on a bill of quantities. One of the metrics that results from these processes may therefore be the 'hit rate' of successful tenders. If the two metrics were compared a performance gap could be highlighted; but what use is this information in isolation? Benchmarking discourages the use of metrics as a means of comparison since they only show the performance gap. If however one of the contractors processes had been identified as best practice, then the metric largely becomes irrelevant. Processes deliver output, in this case the tender sum. Selection of the best possible process will therefore automatically lead to the best output. It is not therefore necessary to concentrate on the metric. A further advantage of concentrating on the process is that unlike the metric, which only indicates the performance gap, it shows why that gap exists. As a direct consequence it shows how the gap can be closed. Benchmarking the process therefore not only highlights the problem but also provides the solution.

Best practice indicates what is achievable

An obvious question to ask in relation to benchmarking is why stop at best practice? In the same way that benchmarking against a company's best internal practices may be encouraging it to fall short of its potential, then so may be the benchmarking of best practice. There is no direct follow-on that external best practice is the best that could possibly be achieved. In addition, in following the best practice of another company, there is an argument that a company is lagging behind instead of leading from the front. The answer to these arguments is twofold. First benchmarking stresses process and within an organisation there may be hundreds of processes and sub-processes. A company is therefore following only in regard to processes, the combination of which will give it business superiority. Second benchmarking best practice shows what is achievable. If a company were to aim higher than best practice there can be no guarantee that it could be achieved.

Provides an environment for change

As outlined earlier, benchmarking requires a shift in business culture. In order to be effectively implemented it requires a change in the way information, internal business processes and competitors are viewed. The problem with this is that with most organisations there is a intrinsic reluctance to change. Advocates of benchmarking argue that the actual process of benchmarking itself can help to overcome this resistance. Because benchmarking is a creative process that investigates how other organisations carry out their business processes, it acts as a catalyst to change. Because benchmarking clearly shows when a performance gap exists, it also helps to motivate employees towards making the change necessary to close the performance gap.

Benchmarking identifies technological breakthrough of other industries

This is often given as one of the advantages of benchmarking and the example usually cited is that of bar codes. Although this was first a technological breakthrough of the grocery industry, bar-codes are now also used in libraries, hospitals, security and identification systems.

Benchmarking allows individuals to broaden their own background and experience

People operating in a particular industry tend to adopt the business culture of that industry. Although this is necessary in order to work effectively within that industry, the business culture may also present restrictions and stifle change. Working with other organisations and industries illustrates that there is more than one way of carrying out any task and that existing methods can almost always be improved.

Benchmarking focuses on the objectives

Benchmarking highlights the performance gap and sets targets aimed at closing that gap. These targets then become a focus or objective for those involved in the benchmarking process. When all staff are focused on the same objectives goals are more likely to be achieved.

The industry best is the most credible goal

One of the problems with implementing successful manufacturing techniques in the construction industry is the claim that the construction industry is different and not conducive to the application of manufacturing techniques, however successful they may have been. A business process however exists regardless of industry and focusing on the best practice of such processes provides a credible goal which can be recognised as achievable. Even for the construction industry!

Step 2 What to benchmark?

This is probably the most difficult part of benchmarking. It is maybe for this reason that the main benchmarking texts do not confer on what in fact should be benchmarked. This text has stressed that one of the fundamentals of benchmarking is process, however not all of the texts agree with this. Peters[11] for example defined three levels of benchmarking; strategic, operational and statistical. Strategic benchmarking deals with benchmarking culture, people, skills and strategy. (Pastore also agreed that benchmarking could take place on strategy [14].) Operational benchmarking deals with benchmarking methods, procedures and the business. (This is also called process benchmarking.) Finally statistical benchmarking which is the numerical or statistical comparisons of company performance.

This idea that more than the process can be benchmarked only succeeds in complicating what is essentially a simple issue. 'Best practice culture' cannot be benchmarked. Even if it could there would be little purpose in doing so, since culture cannot be changed easily, if at all. Culture, including the sub-culture which may exist within an organisation is an intangible asset[15]. In addition there is a close relationship to the type of product a company makes and its corporate culture [16]. As such there is little purpose in benchmarking culture when the products offered are likely to remain different. Making a separate category of benchmarking to deal with purely statistical or numerical comparisons also complicates the issue. As explained earlier, the numerical measure of performance or metric is an output of the process. Comparison of these metrics achieves little other than state that a performance gap exists. As benchmarking is defined as a process of improvement such statistical comparison alone could not be called benchmarking.

There is of course a strong possibility that the technique of benchmarking will develop and it may, at some future date, be feasible that some aspects of corporate strategy or even of people may be benchmarked. However in the context of this text and, the authors believe, in the current and accepted context of benchmarking, the technique embraces only the comparison of products or processes. In the narrower context of construction management the term benchmarking includes the comparison of business processes only, the final product (the building) being too diverse and complex to facilitate meaningful comparison.

How then is a business process defined?

Any business activity can be seen in three stages. There is an input, a process and an output. The combination of these outputs leads to the final product. For the purpose of organisation and administration certain outputs are achieved by grouping them into a business function. Figures 2.4 and 2.5 show how this idea operates in the case of a building contractor. Figure 2.4 shows how the company organises itself into functions, within which processes take place. Figure 2.5 shows in detail the processes and sub-processes that take place within the 'sub-contractor management function'.

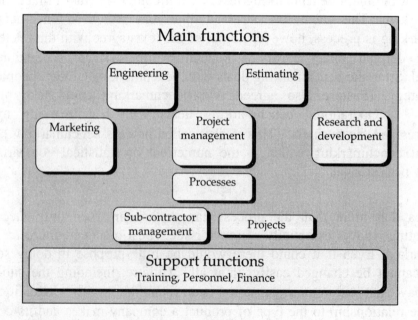

Figure 2.4 The functions of a contracting organisation

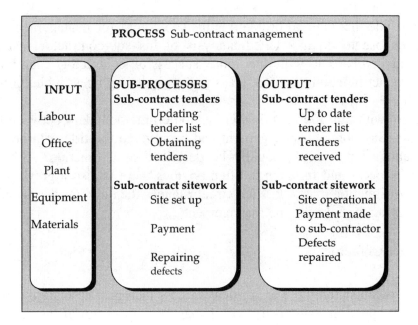

Figure 2.5 Processes contained within the sub-contract main function

In Figure 2.5 project management is the business function and sub-contract management is a process within that. Within this process there are, among others, the sub-processes of sub-contract tendering and sitework. Within the process of tendering there are further sub-processes of keeping an up-to-date tender list and obtaining tenders. These sub-processes take a certain input in terms of resources and in addition produce an output. In this case the output is the completed tender list and the tenders received from the sub-contractors. The sum of all these outputs is the process output, the sum of which is the function output. The sum of all the function outputs is the final product, that is the completed building. This is the basis of benchmarking: to improve the process so that the final product is improved.

The problem that arises however is that as an organisation may have hundreds or even thousands of sub-processes, how to decide which ones should be selected for benchmarking? There are various ways that this can be done and these are outlined below.

Identify the product of the business function

As explained earlier the organisational or administrative divisions within an organisation are its business functions. These can be a good starting point in deciding what to benchmark since business functions tend to be much more visible than business processes. It may for example be known that contractor X has excellent project management. Another contractor, aware his own project management is poor may use this business function as a starting point for his benchmarking exercise. Investigation of the process within that function may lead him to conclude that his real problem is his sub-contractor management. He may therefore choose to do a detailed analysis of his sub-contractor management process and compare it with the more successful contractor. Identifying the superior product function therefore leads to the superior process which produces it.

The problem with this approach is there is no definite boundary between function and process. Sub-contract management had been given the title of process in the example above but it could equally be described as a function. Although the division of process and function is often obvious there is also a grey area where no definite distinction exists. In such circumstances the only answer is experience of the organisation, industry and management.

Critical success factors

Within any organisation there are critical success factors. These are those factors, which, if not operating effectively will damage the operation of the organisation. In the case of the house-builder, marketing may be a critical success factor, whereas in the civil engineering division of a major contracting organisation a critical success factor, may be completion of projects on time. Identification of these factors can be a good starting point for benchmarking. In addition to being critical success factors however items selected for benchmarking must be those an improvement in which will make a significant difference to the overall performance of the company. For example a contractor may know that turnover is a critical success factor but if the company is already operating at capacity an improvement in that area may not substantially benefit the organisation. Critical success factors can be likened to the Pareto rule which states that 80% of the result is based on 20% of the activities. For benchmarking purposes the critical success factors are those 20% activities. It is surprising how few construction managers do actually understand the critical success factors of their own organisation. As a result when organisations get into difficulty managers embark on cost cutting exercises, as this is the only way they can see of improving performance. They believe that reducing cost while maintaining output automatically adds value. In reality such exercises achieve little other than deflating moral and reducing the quality of service offered.

The Xerox questions

To assist in the identification of what should be benchmarked Camp[6] recommended a series of questions, most of which are reproduced below.

What factors are most critical to your business success

In a contracting organisation the following may be critical success factors:

- Turnover

- Tender hit rate
- Ratio of civil engineering to building projects
- Number of houses sold

What areas are causing the most trouble?

In this case the answers may be:

- Management of sub-contractors
- Meeting completion dates

What products or services are provided?

Typically this might be:

- Houses
- Maintenance
- Small building projects

What factors are responsible for customer satisfaction?

If clients of the construction industry were surveyed the following might relate to their satisfaction:

- Completion on time
- Minimum cost
- Guaranteed cost
- Quality
- Health and safety
- Environmentally friendly development
- Architectural expression
- Life cycle cost
- Durability

What are the competitive pressures?

These may be regional, national or international. They may be coming from smaller builders forcing their way into new markets or large contractors looking for smaller projects in a time of recession.

What are the major costs/ Which functions represent the highest percentage of cost?

This may be the head office, permanent staff, manufacturing subsidiaries or marketing.

What functions have the greatest need for improvement?

This could be staff training or ordering of materials.

Which functions have the greatest effect or potential for differentiating the organisation from competitors in the market place.

This is often the most difficult question to answer in that it tries to identify what gives one company the edge over the other.

Ask the customer

Another way of deciding what to benchmark is to ask the customer. In the example used earlier a company may see project management as an area of potential improvement. But who is the customer for this function. Is it the client, the site agent or the sub-contractors? The answer is most likely the latter two. Speaking with these will highlight where the problems of management are and these may become the areas targeted for benchmarking. Customer does not necessarily refer to the final user. In the case of benchmarking customer refers to the customer of the process, that is, the person who is next in the chain and who receives the process output.

It can be seen from the above that there is no definite answer to the question of what processes to benchmark. Any of the above methods or any combination of them could be used to determine the answer. Benchmarking, like many of the techniques in this book, is a soft system. Answers will therefore be highly company specific and dependent of the organisation's culture. The only real answers come from a thorough understanding of the organisation.

A further question which may be asked when deciding what to benchmark is what is the level of detail that is required. As shown above a business process may be broken into many sub-processes and these sub-processes may be broken down still further. The level of detail selected is entirely up to the individual organisation and will vary according to the size of the organisation and the amount of resources they wish to dedicate to the benchmarking activity.

Step 3: Which companies to benchmark?

The answer to this question is simple yet very difficult. The simple part is that the comparison or benchmark company is one who's processes are best practice. The difficulty is how to find out who these companies are. In some functions, such as marketing, finding out who is the industry best can be quite straightforward. In the case of house building particularly, information on sales, advertising and brochures can be collected easily. In addition sites can be visited and show-homes and site marketing examined. If resources are available house purchasers can be surveyed and a marketing consultant employed to analyse the results. Once all available information is collected then a judgement of best practice can be made.

In the case of a civil engineering division of a major contractor the process of identifying best practice becomes much more difficult. The product, not being something that the general public purchase, is further removed from the market place. The number of finished products is limited, there are no brochures and little data about projects in the public domain. How then can the information regarding best practice be obtained?

The only way to establish leading companies in a specific function or process is through a collection of data. Most of the benchmarking texts give a list of sources of information to locate best practice companies and the major sources are reproduced below.

- Special awards and citations
- Media attention
- Professional associations
- Independent reports
- Word of mouth
- Consultants
- Benchmarking networks
- Internal information
- Company sources
- Library
- Experts and studies
- Questionnaire by mail and telephone
- Direct site visits
- Focus groups
- Special interest groups

- Employees, customers and suppliers
- Foreign data sources
- Academic institutions
- Investment analysis
- The internet
- On line data base
- Journals

Step 4: Establishing the performance gap

Once best practice has been identified the company who owns it needs to be approached to be a benchmarking partner. To most construction management professionals this presents the biggest problem of benchmarking, in that there is a reluctance to approach other companies or competitors. Likewise there is also a reluctance on the part of the company approached, to give information which will be used as a basis for improving their competitors. As outlined by Spendolini[7] most people employed in private enterprise see it as their role to beat the competition, not train them.

This attitude is understandable; after all it is based on years of industrial practice. However in reading all the sections of this book one item should by now be clear. That is that all of the techniques presented, be they total quality management or value management, require some sort of cultural shift in the way the construction industry operates. Markets in all industries have changed beyond recognition in the last fifteen years. Markets are now global and customer focused. The construction industry has been slow to recognise this but external competition means it cannot ignore it any longer. If the industry is to survive it must change. A part of this change will undoubtedly demand some degree of co-operation between companies.

There are no definite rules on how a company should be approached with a request to become a benchmarking partner; this is a question for the individual company concerned. However the critical factor is that it is a partnership. Both parties should expect to gain something from the experience, even if it is how to conduct a benchmarking exercise.

Having identified best practice and entered into a benchmarking partnering agreement the next step is to fully document the process that is under study. There are many methods of doing this such as flow charts or process diagrams. Space does not allow a detailed examination of them here, but all good management texts will usually explain these techniques thoroughly.

In earlier stages of this chapter use of the metric or quantification of the process output was discouraged. However now that a thorough examination of the process is documented the metric can be used. In some instances use of the metric can be highly relevant, examples being tender hit rates, projects completed on time or claims made by sub-contractors. In some instances however it may be sufficient to simply document the process and compare the processes under study. The movement of a tender within an organisation is a good example. Where a metric is used it can take several forms: ratio, cost per contract, cost as a percentage of turnover. Whichever method is used the company will form one of the three judgements about its own performance of a particular process in relation to that of the benchmarked company:

Parity: that both companies have equal performance.

Positive performance gap: that the company under study is in fact performing better than best practice.

Negative performance gap: that the company under study is performing below the level of best practice.

Step 5: Set targets

The most likely outcome will be that company performance is below that of the benchmarked company and that there is a negative performance gap. In such instances the company will need to devise how the gap will be closed. Unless the gap is small and can be closed in one company incentive then it will be necessary to set a target aimed at achieving the best practice in a series of stages.

It is not the intention of this text to give a detailed account of the procedures for goal setting within the organisation, since these can be found in any good management text. Suffice to say the target must be planned, realistic and achievable. It must be communicated to staff and require their participation.

Steps 6, 7 and 8: Implement-Monitor-Set new target

The next stages of the benchmarking process require that the plan to achieve the first target be implemented and monitored, followed by the setting of a new target. Once again procedures for doing this are not included here since they can be found in most management texts. In addition most companies will have their own procedure for implementing plans and monitoring progress. Lack of inclusion here is not intended to reduce the significance of these stages of the benchmarking process. They are of course vital to its success. A plan that is made and not implemented is worse than no plan at all.

Step 9: Best practice achieved

Best practice achieved means the company has achieved parity with the benchmarked company. However the benchmarking process does not end here. As outlined earlier benchmarking is a continuous process. Like TQM it is something a company has, not something a company does once in a while. In order to achieve superiority a company needs best practice in all of its major processes. It can only achieve this by constantly looking for further improvements.

The first section of this chapter has looked at the steps to be taken in a benchmarking study. The second section deals with items relevant to the implementation of the study.

The benchmarking team

Benchmarking, unlike value management, is not necessarily a team activity; it could be carried out by an individual. However in the interest of sharing the workload and also as an aid to acceptance and communication of the benchmarking, then a team is the recommended approach. There are various formats of a benchmarking team.

Work groups

These are groups that already exist and may comprise members of an existing department or group.

Interdepartmental groups

These are formed by bringing together people with particular expertise in the area required. These teams will generally disband after the benchmarking study is complete.

Ad hoc teams

These are more flexible and made up of any person interested in joining the benchmarking group. Group members may come and go. Benchmarking groups of this type will not work well unless the company is a mature benchmarking company such as the Xerox Corporation.

Benchmarking Code of Conduct

The American Productivity and Quality Centre's International Benchmarking Clearing House and the Strategic Planning Institute Council of Benchmarking have developed a code of practice for benchmarking[4]. It contains nine basic principles as outlined below.

Legality

This principle excludes anything contrary to restraint of trade, such as bid rigging.

Exchange

Benchmarkers should not ask another company for information that they themselves would be unwilling to share.

Confidentiality

Nothing learned about a benchmarking partner from a benchmarking study should be shared with anyone else.

Use

Information gained from benchmarking studies should not be used for any purposes other than process improvement. Advertising is the most obvious example.

First party contact

The first contact at the company to be benchmarked should be the person responsible for benchmarking, not the person is charge of the process.

Third party contact

Company contacts should not be shared with other companies without first gaining permission.

Preparation

Benchmarking partners should not be contacted until all preparatory work is complete.

Completion

Commitment to benchmarking partners should not be made unless it can be followed through to the completion of the study.

Understanding and action

Before embarking on a benchmarking study both companies should understand the process and have made their intentions clear, particularly about the use of information.

Legal considerations

The following are the major legal considerations of benchmarking[17].

- Chapter 1 of Part III of the Rome Treaty, particularly articles 85 (Sub 1,2,3) and Regulation 17 (article 4). These agreements apply to all EU countries and state that agreements of price fixing, sharing markers, discrimination against third parties to their competitive disadvantage or imposing territorial restrictions which partition the common market are prohibitive.

Domestic legislation includes:

- The Restrictive Trade Practices Act 1976
- The Fair Trading Act 1973
- The competition Act 1980

Benchmarking: the major issues

It is not the intention of this book to 'sell' the technique of benchmarking. Rather it is to give an objective appraisal of how the technique can operate in the construction industry. In that vein the following is a summary of current thinking which those concerned with more academic aspects of benchmarking may find interesting.

The relationship between benchmarking and TQM

Most benchmarking texts agree that benchmarking works best in companies with a culture of total quality management. Jackson et al[18] argued that without a total quality control programme benchmarking is a waste of time and money. Whether or not this is true, particularly in the construction industry, is really not yet known.

The place of innovation in benchmarking

One of the criticisms of benchmarking is that because it follows as opposed to leads it is not an innovative technique. Watson[4] however saw benchmarking as only part of the equation and that superiority lies in quality beyond the competition, technology before the competition and cost below the competition. The authors agree with this. Benchmarking is not, or was not intended as, a tool of innovation, but of process improvement. Bendell et al[12] however in contrast to this view saw that the technique could be innovative and that a forth type of benchmarking, customer benchmarking, goes beyond what the customer expects or requires. In this way it is an innovative process and not only one of process improvement.

The problem of using metrics

A paper in World Wastes[19] used the metrics of waste services at several major American cities. This is a classic example of why metrics should be avoided when benchmarking. The paper described how benchmarking was used 'to measure one service delivery system against another,' and compared the number of refuse collections along with costs of waste services. However as this process neither selected best practice nor aimed at process improvement, it could not be described as benchmarking. It was a simple comparison of quantifiables.

Although theoretically there is clearly logic in delaying the use of metrics until the business process is documented the reality is much more difficult. Managers will want to see at least an estimate of the size of performance gap before they allocate resources aimed at closing it. Arguments about the need for a cultural shift will have little effect on the manager with limited resources. The answer here is that metrics can be used earlier in the benchmarking process if there is no alternative. However they should always be used with caution, and, on documentation of the process, be recalculated.

Business co-operation

For many managers this is the most difficult aspect of benchmarking. There are two basic arguments in favour of business co-operation. First the essence of benchmarking in construction lies in the process, which is not as sensitive as the product. Second and strongly in keeping with the theme of this book, there is a desperate need for a cultural shift in the construction industry. Benchmarking, according to Watson[4], is about a coherent *national challenge* for continuous improvement. Many companies already recognise that future success depends on the global market place and the development of national productivity. Without co-operation there can be no benchmarking. Any company who cannot tolerate this concept will not therefore be able to carry out the benchmarking process.

How far to go?

In the search for best practice how far should a company go? Should, as recommended by Grinyer and Smith[20], the search be a global one. The answer to this must realistically be no. It is not feasible for a medium sized building contractor to search the world as a means of finding improved processes. Very few organisations are capable of such an extensive and exhaustive search. The benchmarking partner must clearly be a comparable size and status company that offers processes with a high degree of applicability and relevance.

Japanese Kaizen

Benchmarking is, in some texts and articles, compared to Japanese *kaizen*[21]. This however is not a correct comparison since *kaizen* is a much broader concept that does not rely solely on a system of external comparison. *Kaizen* is examined in greater detail in the chapter on total quality management.

Disadvantages of benchmarking

Lenkus[22] looked at some of the disadvantages of benchmarking namely that it is complex, cumbersome, time consuming and expensive. He also found problems with securing benchmarking partners. Although he could see the advantages of benchmarking metrics he saw that benchmarking processes was difficult. He also believes that managers invariably want to see metrics before they will commit resources to process improvement. Other disadvantages included the length of time taken and the fact that technological change may move faster than the time it takes to carry out a benchmarking study.

Current research

Benchmarking, particularly in construction, is relatively new. As such there has been little research carried out on the technique. The following sections deals with the findings of the work that has been carried out to date.

Benchmarking has been carried out on the pre-planning phase of construction projects[23]. This work looked at four major sub-processes in the pre-planning process of 62 construction projects. These were organise, select project alternatives, develop a project definition package and decide whether to proceed with the project. The work collected data on these processes and analysed the data. There are several contentious issues with this work. First is that the process and sub-processes examined were based on a theoretical model against which the 62 projects were compared. However as the model is purely theoretical no judgement about project performance can be made in relation to it, since if the model were incorrect then so are all the resulting judgement. The second question which arises from the research is whether a construction project and its processes are in fact capable of being benchmarked. The number of design and constructing firms involved make it an incredibly complex process [24]. A construction management team is in effect a temporary organisation with one off processes established on an ad-hoc basis. Given a limited design and construction span and the fact that many organisations input into the processes, how could best practice ever be established? Even if a project's processes were identified as best practice then the introduction of a new client, consultant or contractor would surely make repetition of them impossible. This raises a very serious question for the future of research into benchmarking in construction, namely, can the project be benchmarked or can benchmarking only apply to the individual organisations that exist within the project.

Another interesting aspect of benchmarking research is the relationship of the technique to the culture within which it exists. Watson[4] recognised that given that the future of benchmarking lies in its global application then benchmarking cannot be viewed outside the contexts of international trade and culture. He sees a need in the future of benchmarking to bridge those cultural gaps. However national differences are not the only culture that may affect benchmarking. Company or corporate culture may also be related to it. Companies have three types of performance behaviour[14] which are basic, innovative and competitive. In the innovative sphere a company takes a risk and hopes the customer will like it. The Sony walkman is a good example of this. With competitive behaviour, competition is based on direct comparison of product features among product alternatives. Unlike in innovative behaviour, the requirements of the customer are not guessed at but are already known. Basic behaviour satisfies the customers' lowest level of expectation and provides only the product features that customers assume will be provided. There can be no question that these corporate cultures exist in the manufacturing field but do they also exist in construction? If not then is there any value in a construction company benchmarking outside the industry? Do the barriers provided by corporate culture simply become too great?

Spendolini[7] produced an excellent analogy of how cultural differences have an effect on benchmarking. It was what he called thinking out of the box. The further out of the box the company is prepared to go, the greater the cultural differences that are encountered. The rewards however may be greater. This is an interesting idea, however there is no real evidence that it is true, or even possible to achieve.

The work of the Construct IT Centre was mentioned earlier[9]. In this work 11 construction companies took part in the benchmarking of IT use in construction site processes. The companies nominated their best project from the point of view of IT use. This work used an engineering company as the benchmark as its business processes were similar to that of the construction industry. The report made several recommendations for effective management of the site process. These included a strategy which elevates the importance of IT, a re-examination of business processes by construction companies, greater investment in IT and greater education of the workforce in the use of IT.

A case study

In his book Glen Peters[13] examined the benchmarking of customer service. This has traditionally been an area of weakness in the construction industry and as an example the authors carried out a simple exercise of benchmarking the approach of an average house-builder to customer service, compared with the approach taken by a leading car manufacturer. Car manufacturers were chosen because the marketing of cars, the product identification and also the customer service provided are visibly superior to anything offered by the house-builder, yet both are dealing with expensive commodities that are bought fairly infrequently. As explained earlier, since benchmarking concentrates on process which cuts across industry barriers it is valid to benchmark the house-builders customer service against that of the car manufacturer. The survey questions are taken from the benchmarking example survey contained in Peters[11] book and is used with his permission. The results are shown in Table 2.2.

Obviously no definite conclusions can be drawn from such a small survey but, as expected, the car manufacturer was more customer-focused than the house-builder. Not only were they more aware of the costs of gaining and losing a customer but they also had a distinct complaint taskforce that recorded every complaint made. In addition, and unlike the house-builder, all of their front-line staff had received customer-care training. An interesting aside to this is that a quantity surveying practice was also asked to complete the survey. They returned it uncompleted, claiming they could not understand the questions!

Conclusion

One of the items that the authors found of greatest interest when writing this chapter was a paper by Yoshimori[25]. In this he looked at how senior managers in different countries regarded the ownership of their companies. In the UK 70.5% of senior managers saw shareholders as their first priority. In Japan only 2.9% took this view. However in reply to whether priority should be given to all stakeholders, which includes employees 29.5% of British managers responded positively compared to 97.1% of Japanese.

Camp's[6] book which is the first text written on benchmarking outlines the historical development of benchmarking in Xerox. In this he clearly states that the overall aim of the Xerox programme was to achieve leadership through quality and that this was achieved through three components of benchmarking, employee involvement and the quality process. However most of the texts that followed Camp failed to mention this, concentrating on the technique of benchmarking as a stand alone technique. Nowhere is this highlighted better than in a paper by Jackson et al[18] which makes an appraisal of the many benchmarking texts currently available. The paper gives a rating to the coverage of certain major

Table 2.2 Benchmarking customer service in the construction and automotive industries

Question	Car company	House-builder
How do you identify who your customer are	Database Focus groups Quantitative written research Exist interviews	Database Current active accounts Planning leads
How do you gather information about potential customers	Direct marketing Inserts Through dealers	Written Telephone surveys
How often do you re-examine the information you have about potential customers	Annually	6-12 months
What percentage of your customers are identified	Don't know	70%
What percentage of your overall customer/market research budget do you spend on identifying customers	100%	10%
What percentage of complaints are recorded	100%	40%
How do you store information about complaints	Computerised system organisation wide	Manual system on each site
Do you calculate the costs of losing a customer	Yes	No
Do you calculate the costs of gaining a customer	Yes	No
Have you identified the areas that produce the most complaints	Yes	No
What percentage of front-line staff receive training	100%	80%
What do you do to analyse complaints	Complaints task force	Front-line employees are asked to recommend change
Do you tell customers what has happened as a result of their complaint	Yes	Yes and no
Do you offer some sort of compensation to a complaining customer	Yes	Generally no

topics including total quality management process. However no topic is included on employee involvement, as this aspect of benchmarking appears to have diminished in importance. Many of the books on benchmarking hold up the Japanese as a great example of what benchmarking can achieve. Although this may be true, it may equally be the case that Japanese success is partly a function of employee involvement within the organisation. This aspect of benchmarking cannot be entirely overlooked. Even at the 'source' of benchmarking at Xerox employee involvement was seen as one of the essentials of a three part programme aimed at successful business. This however seems to be forgotten in the development of benchmarking. If it is continually overlooked then it is the authors' view that benchmarking will fail and simply join the ranks of what was termed the flurry of acronymic assaults that have accosted the business world[4].

---

[1] Watman M. (1964) *Encyclopaedia of Athletics*. Robert and Hale, London.

[2] Temple C. (1990) *Marathon, cross country and road running*. Stanley Paul, London.

[3] Drucker P. F. (1994) *The practice of management*. Butterworth Heinmann, Oxford.

[4] Watson G. H. (1993) *Strategic benchmarking. How to rate your company against the world's best*. John Wiley, New York.

[5] Codling S. (1992) *Best Practice benchmarking*. The management guide to successful implementation. Industrial newsletters limited.

[6] Camp R. C. (1989) *Benchmarking: The search for industry best practices that lead to superior performance*. ASQC Quality Press, Milwaukee, Wisconsin.

[7] Spendolini M. J. (1992) *The Benchmarking book*. American Management Association, New York.

[8] Department of Trade and industry. (1996) *Inside UK Enterprise. Managing in the 90's*. Further information can be obtained from Status Meetings Limited, Festival Hall, Petersfield, Hampshire. GU31 4JW. Tel:01730 235015 Fax: 268865.

[9] Construct IT Centre of Excellence (1996) *Benchmarking Best Practice Report. Construction Site Processes*. University of Salford.

[10] The Building Research Establishment LINK IDAC (1996) *Project: Benchmarking for construction*. In Innovative Manufacturing Initiative Report Benchmarking Theme Day, 25 July 1996, London. Published by the Engineering and Physical Science Research Council, Swindon.

[11] Karlof B. & Ostblom S. (1993) *Benchmarking: A signpost to excellence in quality and productivity*. John Wiley.

[12] Blendell T., Boulter L., Kelly J. (1993) *Benchmarking for competitive advantage*. Financial Times Pitman Publishing, London.

[13] Peters G. (1994) *Benchmarking customer service*. Financial Times Pitman Publishing, London.

[14] Pastore R. (1995) Benchmarking comes of age. *CIO* **9**, (3), 30-36.

[15] Gray S. (1995) Cultural perspectives on the measurement of corporate success. *European management journal*. September, **13** (3), 269-275.

[16] Kono T. (1994) Changing a Company's strategy and culture. *Long range planning*. October, **27** (5), 85-97.

[17] Coonen R. (1995) Benchmarking: A continuous improvement process. *Health and safety practitioner* October, 18-21.

[18] Jackson A. E., Safford R. R., Swart W. W. (1994) Roadmap to current benchmarking literature *Journal of Management in Engineering*, **10** (6), 60-67.

[19] Slovin J. (1995) Cities turning to benchmarking to stay competitive *World Wastes* **38** (10), 12-14.

[20] Grinyer M. and Goldsmith H. (1995) The role of benchmarking in re-engineering. *Management Services*, October, **30** (10), 18-19.

[21] Anon. (1995) Benchmarking Kaizen *Manufacturing Engineering*, November, **115** (5), 24.

[22] Lenkus D. (1995) Benchmarking support is split. *Business Insurance*, **29** (39), 1, 29.

[23] Hamilton M.R. and Gibson G.E. (1996) Benchmarking pre-project planning effort. *Journal of management in engineering*, March/April, **12** (2), 25-33.

[24] Mohamed, S. (1996) Benchmarking, Best practice- and all that. *Third International Conference on Lean Construction*. University of New Mexico Albuquerque, 18-20 October.

[25] Yoshimori M. (1995) Whose company is it? The concept of the Corporation in Japan and the West. *Long Range Planning*, **28** (4), 33-34.

3 Reengineering

Introduction

Despite being a relatively new discipline there is already some ambiguity with respect to what is meant by 'reengineering'. As a starter there are variations in the spelling. Some commentators favour the hyphenated version 're-engineering', others use the non hyphenated version 'reengineering'. Hammer, who is usually credited with the introduction of the concept, favours the non hyphenated version and we have opted for this spelling in this text. Most common use dictionaries, such as Webster's do not recognise the term in either form.

Confusion can also arise on whether to use 'reengineering' as the generic term which applies to all reengineering activities or to use the more specific term 'business process reengineering' (BPR). Because most references on the subject are from a business background the term BPR is frequently used as the generic. We have however opted to use reengineering as the generic term (as does Hammer) because the term construction process reengineering (CPR) has recently been coined[1] and, for the purposes of this text, we need to be able to distinguish between BPR and CPR.

Much is currently being claimed about the power of reengineering. The flyleaf of Hammer and Champy's text 'Reengineering the corporation'[2], makes the claim that *'In Reengineering the corporation,* Michael Hammer and James Champy do for modern business what Adam Smith did for the industrial revolution in the *Wealth of Nations* two centuries ago - they reinvent the nature of work to create the single best hope for the competitive turnaround of business'.

We have throughout the text attempted to be as detached and objective as possible. However, of all of the concepts in this book, reengineering has been the most difficult for us to maintain our level of detachment. This is, in the main, due to the fact that reengineering is seen by many of its proponents as a revolutionary rather than an evolutionary concept, and because of this much of the language used to describe reengineering has an emotive flavour. For example, Hammer's often quoted saying 'don't automate, obliterate', typifies the revolutionary hype of reengineering parlance. Many texts on the subject make claims such as 'radical change', 'dramatic results'. Hammer and Champy use both these expressions in defining reengineering as, 'a radical redesign of business processes in order to achieve dramatic improvements in their performance'. Morris and Brandon[3] take the extreme view that 'You can choose to reengineer or you can choose to go out of business.' It has been difficult for us to present a cool clinical exposition of the topic in the light of comments such as those by Morris and Brandon. There is however a counterpoint to this aggressive promotion of reengineering in the work of COBRA[4] (Constraints and opportunities in business restructuring - an analysis) which is an initiative of the European Commission. Professor Coulson-Thomas, leader and co-ordinator of the COBRA project describes the back drop to the project as follows:

'Creativity and imagination are at a premium. According to its theory and rhetoric Business Process Re-engineering (BPR) is concerned with step change rather than incremental improvement; revolution not evolution. However, while BPR is being used to 'improve' existing situations, i.e. cut costs, reduce throughput times and squeeze more out of people, what does it contribute to radical change and innovation?'

What we have tried to do for the reader is to take the middle ground, as far as this is possible, in presenting an overview of reengineering and exploring the implications of reengineering for the construction industry.

We should say at the outset that we are more inclined to the Coulson-Thomas's view that reengineering is not the only way to achieve radical and fundamental change and that creative thinking, benchmarking, cultural change and innovation can be undertaken independent of reengineering. Reengineering is however an important management concept which has achieved quite dramatic and tangible improvements for large business corporations and, as such, merits serious consideration.

Origins of reengineering

Although the emergence of reengineering is usually attributed to the publication in 1990 of Hammer's Harvard Business Review Article[5] 'Reengineering work: Don't automate, obliterate', the origins of the reengineering concept can be traced back to the 1940's to the work of the Tavistock Institute in the UK where the 'social technical systems approach' was applied to the British coal industry[6]. The essence of the social technical systems approach (STS) is that technical and social systems should harmonise to achieve an optimal overall system. The STS approach was translated in the 1980's[7] into Overhead Value Analysis (OVA) which emphasises that attention must be centred on the work being done rather than on the people doing the work. According to Rigby[6], OVA was in fact, the precursor to reengineering.

The underlying theme of the STS, the OVA and the reengineering approach is that work processes need to be redesigned. The implications of reengineering, according to Hammer, is not that we design inefficient processes to begin with but that these processes have been overtaken by time and have not evolved or kept pace with advances in technology.

'Over time, corporations have developed elaborate ways to process work. Nobody has ever stepped back and taken a look at the entire system. Today, if most companies were starting from scratch, they would invent themselves in totally different ways.' The added dimension which reengineering gives to STS and OVA is to the concept of 'discontinuous improvement'. The imperative for reengineering is, to achieve a quantum leap forward rather than small continuous gains[8].

Reengineering in a construction industry context

In considering reengineering in a construction industry context, the first thing which needs to be determined is whether or not the construction industry is a special case, i.e., should it be treated differently from all business processes? There are arguments for and against treating construction organisations as somehow different from other business organisations. Certainly the construction industry has a set of characteristics and a product which is uniquely its own. However delivering a product with a unique set of characteristics is not a convincing argument for special status. The automotive industry and the white goods industries produce quite different products, one producing cars which must be able to perform in the open air and the other domestic appliances which operate in an internal environment, however no one would suggest they are not both part of the business process.

The argument is often made that the construction industry is highly compartmentalised, highly fragmented, under capitalised, and operates on a single project-by-project basis, and so on. The more that we examine the special pleadings that the construction industry is different from all other business process, the more convincing becomes the argument that the industry is a prime candidate to be reengineered. To advance this argument further, the starting off point for the reengineering process is to take a new or green field approach, a feeling of wiping the slate clean. There is no better way for the construction industry to do this than to regard itself as a member of the international business community, on a similar footing to the MacDonald's, Xerox's and Toyota's of the world. Although, in general, the scale of the construction organisations is smaller than that of other multi-national business organisations, the output of the industry is usually in the order of 8 to 10% of the Gross Domestic Product (GDP) of a westernised economy, with a 10% improvement in construction performance representing a 2.5% increase in GDP.

The view which we would like readers to consider, in the context of reengineering, is that the construction industry should be seen as *part of*, not *apart from* the general business community. This is not meant to infer that the construction industry has everything to learn and nothing to contribute to the business community at large.

The current trend in business process reengineering applications is for these to be internally focused within the organisation: 'the BPR application becomes exclusively an 'in-house' operation, in which, decisions pertaining to meeting customer requirements, investing in technology and organising resources are steered and, if necessary, manipulated by the organisation, with minimum dependence upon and interference with external factors [9].' Whether or not reengineering can be applied to an industry sector as a whole rather than to individual organisations is a challenge which has still be addressed. 'An industry, such as the construction industry, represents a major challenge to BPR. This is mainly due to the complex business relationships which dominate the industry plus the key role external factors play in how construction industry organisations conduct their business[9].'

Our point is that although the construction industry does have peculiar characteristics, it is still part of the business community. If the construction industry can apply reengineering not only at the level of the firm but also at the level of a complete industry sector then this will be a major contribution to the advancement of the reengineering cause and would be a considerable achievement for the construction industry. Whether or not this can be achieved is another matter, which will be considered later.

The goals of reengineering

For many years the construction industry has focused on delivering buildings on time, within budget and to a specified quality. These goals are quite compatible with the goals of reengineering which, are 'to achieve dramatic improvements in critical contemporary measures of performance, such as cost, quality, service and speed'[2], an area that the construction industry is only recently coming to grips with. By and large the goals of reengineering are no novelty to the construction industry, what is novel however is the green field methodology which is the essence of the reengineering concept. Hammer and Champy[2] distil reengineering into four key words or characteristics: fundamental; radical; dramatic and process. The implication of these key words pose some interesting questions for the construction industry.

The first characteristic of the reengineering approach is the requirement to take a fundamental look at what is the core business of a company. In essence the question being asked is 'why are we here?' 'why do we operate the way that we do?' The expected response from a business organisation to the question 'why are we here?' would be 'to service our customers' or 'to meet our customers' expectations'. Regrettably the construction industry does not have a good track record with respect to identifying the customer as the *raison d'ître* for the organisation's existence. A recent survey of construction companies in Australia demonstrated that although many companies profess an interest in obtaining feedback on customer involvement, only 5% of those surveyed conducted a formal survey to gauge satisfaction levels[10]. Reengineering demands that nothing be taken for granted, particularly the client. The status quo is not sacrosanct. The primary reengineering objective, when applied to the construction industry as a business process, is to exert pressure on the players in the industry to truly understand the nature of their business.

The second characteristic of reengineering is its emphasis on taking a radical approach. The word 'radical' is often taken to mean 'new' or 'novel'. It is worth remembering that the derivation of radical is from the Latin word 'radix' meaning root. Webster's Dictionary defines radical as 'relating to the origin', 'marked by a considerable departure from the usual or traditional'. (Hammer[2] makes particular emphasis of this.) A radical approach in the construction industry would be to ask the question 'how can we redesign the process so that waste is eliminated?' rather than 'how can we minimise materials wastage on site?' Or at a macro level, consideration could be given to using the principles and tools of 'lean construction' which encompasses the whole gamut of decreasing waste in both the design and construction process.

It is worth stressing that although reengineering demands a radical approach to problem definition and problem solving, this does not mean that existing management tools should not be deployed. Although the essence of reengineering is the novelty of its approach, the use of management tools such as total quality management, benchmarking, concurrent construction and lean construction are essential components of the reengineering approach.

There seems to be general agreement that Hammer and Champy's third characteristic, 'dramatic' typifies the successful application of reengineering. The effects of a successful reengineering strategy should be discontinuous improvement, i.e., a quantum leap forward. The notion of a quantum leap may seem to infer that only companies who are performing badly will reap the full benefits of reengineering. Hammer and Champy dispute this and identify three kinds of companies that can benefit from reengineering. These are companies who are in deep trouble and have no choice but to undertake reengineering in order to survive. The second company type is one which is not yet in trouble but whose management has the foresight to see that trouble is brewing ahead. The third company is one which is a leader in its field which is not under threat either now or in the foreseeable future but is keen and aggressive. Thus no sector of the business community is absolved from involvement in the dramatic impact of reengineering.

One question which does however spring to mind in terms of the dramatics of reengineering, is 'how can a company exist in a state of perpetual discontinuous improvement?' In other words once having achieved a quantum leap in business performance through reengineering will the law of diminishing returns set in so that additional applications of reengineering result in ever-decreasing gains? This conundrum has been addressed by Morris and Brandon[3] in what they describe as the 'change paradigm.'

'The change paradigm is a conceptual environment. Once a company begins operating within this paradigm, the process of reengineering never ceases. It becomes constant but incremental, as the company evolves toward better quality and efficiency. This represents a new business operation cycle.'

Thus the change paradigm addresses the conundrum of the potentially self-defeating nature of reengineering by recognising that in the future the construction industry will compete in an environment which is in a dynamic state of change where the only constant is change itself.

The fourth and final key word in Hammer and Champy's definition of reengineering is 'process'. The underlying contention is that most businesses are not process orientated. The primary concern of reengineering is in redesigning a process, not in redesigning the organisational structure of departments or units. The distinguishing characteristic which reengineering brings to this redesign is that it should be fundamental, radical and dramatic. 'The majority of the organisations who use the term business process reengineering are in reality engaged in process redesign. For most companies this approach represents radical change. However, it is not what Michael Hammer and James Champy meant by the word "radical" [11] .' It is important to differentiate between process improvement, process redesign and process reengineering (Figure 3.1)[11].

Process improvement involves a minor degree of change with a corresponding low degree of risk together with a low expectation of improved results. Process redesign is the middle ground with a moderate degree of risk and consequently higher expectations of improvement. Process reengineering is at the end of the spectrum with a high risk, high gain set of expectations.

These three levels have alternatively been described as process tidying, process tinkering and process reengineering[12]. Process tidying is described as a method by which existing flows of people, information and materials are mapped and streamlined by identifying opportunities for eliminating dead-ends, *ad hoc* activities and duplication. Process tinkering is described as a method by which organisations find short-cuts in their processes or identify more user friendly ways of doing work. In general, process tinkering does not seek to change the overall process nor does it seek to move constraints, whereas process reengineering is the method by which physical or mental recipe-induced constraints are eliminated from the organisation and re-established in a way which meets the goals of the organisation.

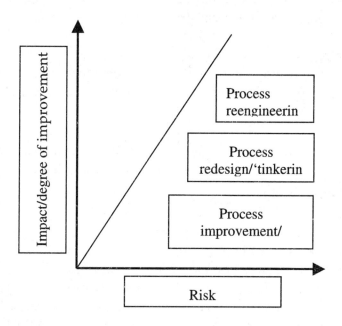

Figure 3.1 Differentiating between process improvement, process redesign and process reengineering

Reengineering methodology

The initiation stage

Reengineering is about cultural change. It is concerned with decisions of a strategic and often political nature. It is about changing attitudes and behaviour and points of reference. It follows therefore that tactically, reengineering must be initiated at the highest level within a company. It is a top down approach. Hammer and Champy[2] observe that the following roles tend to emerge in companies who have implemented reengineering:

The leader: In a construction company this would typically be the managing director or some one with the equivalent authority who would initiate the reengineering concept.

The process owner: Reengineering will be initiated at head office level, not at site or project level. The process owner is therefore likely to be a director or very senior manager based at head office.

The reengineering team: The team would consist of a multi-disciplinary group who are committed to the concept of reengineering with skills and competencies ranging from construction management to IT.

Steering committee: This would be a small group of senior managers who are highly regarded within the construction company.

Reengineering 'czar': In Hammer and Champy's words 'an individual responsible for developing reengineering techniques and tools within the company and for achieving synergy across the company's separate reengineering projects.'

DuBrin[13] places particular stress on the importance of the process-owner or case-manager as a person and also the importance of this person having the authority to leap over departmental walls.

The examples of reengineering very largely emanate from a factory floor production environment, which has organisational groupings which are not necessarily akin to the construction industry. It is worth bearing this in mind and, whilst there is nothing in Hammer and Champy's definition of reengineering roles which would be an anathema to the construction industry, we must be aware that in the construction industry we are dealing with a bespoke product derived from an organisational structure which commonly has a complete separation of the client, designers, constructors and users. Thus the process owner in the construction industry reengineering context must, if we use DuBrin's analogy, have the ability to leap over a large number of extremely high walls.

The planning stage

There seems to be a general agreement amongst commentators and practitioners of reengineering that the benefits of reengineering come from the holistic nature of the concept. In reengineering the emphasis is on horizontal rather than vertical integration, with the breaking down of barriers, the removal of specialist functions and basically a redrawing of existing boundaries. The overbridging strategic methodology recommended by most practitioners has its origins in the systems approach.

The conventional systems approach to planning a project is firstly to identify the system in general terms by establishing the system boundaries and then, and only then proceed to the specifics. This approach is illustrated in Table 3.1 after Armstrong[14]:

Table 3.1 The conventional systems approach

Conceptual	Operational
(Start here) Ultimate objectives	Indicators of success
Alternative strategies	Operational program *(Finish here)*

In a classical systems approach each step should have a separate time period with time for reflection between each step. Each step should be considered without reference to the steps which come later. The sequence of steps is then applied iteratively.

The application of systems thinking can be seen in Table 3.2 which is loosely based on The Texas Instruments diagram of the components of business process reengineering[11].

Step one 'identifying objectives' means starting at the highest conceptual level and determining 'what the business needs to be' by resolving the stakeholder interests of 'what the customer wants' and 'what are the business objectives'. This first step should be completely resolved before proceeding to the next step, 'develop indicators of success'. Developing indicators of success is an important element of the systems approach, this can only be done through a thorough understanding of what the 'business is'. Once the first two steps have been completed decisions can then be made in terms of selecting the alternative strategies of either business process reengineering or continuous improvement, thus leading to step four, the development of programs. The whole process is dynamic with the steps being repeated, in sequence, as required.

Table 3.2 The application of systems thinking

Steps	Procedures
Identify objectives	Resolve customer needs and business objectives
Indicators of success	Indicators of success are developed through an understanding of the business process
Consider alternative strategies	Select either: business process reengineering or continuous improvement
Develop and select programs	If BPR selected then embark on a program of fundamental change. If continuous improvement selected then embark on a program of fine tuning

The COBRA project[15] developed a six stage business process reengineering methodology which has similar characteristics to the Texas Instruments model. It also includes the iterative loop of the systems approach.

The implementation stage

As we will discuss later in this chapter, there are pitfalls associate with the implementation of reengineering, however, for the moment, we will concentrate on the factors which are likely to contribute to successful implementation. It is worth however bearing in mind the contention 'that radical change is accompanied by substantial risk[11]'. In reengineering the risks are heightened by the length of time needed to complete the implementation stage and to get results. The need to 'get on with it' must always to be tempered by the need to prove, on a step by step basis, that the implementation process is working.

Coulson-Thomas[16] identifies 15 success factors in implementing reengineering based on his own experience and observations. Many of these factors relate to the 'people' component of reengineering. In his view reengineering is, in essence, the effective management of fundamental change. He cites success factors such as mutual trust between senior managers and change teams as being critically important: in other words, reengineering is largely concerned with 'feelings, attitudes, values, behaviour, commitment and personal qualities such as being open minded'.

The importance of the human/communication factor is also reflected in the following key success factors identified by Mohamed and Yates[17] as pre-requisites in implementing reengineering in a construction industry context. (Readers with a particular interest in how reengineering might be applied to a specific construction work package are recommended to read this paper in full.)

- Strong commitment by designers, consultants, and contractors to make a major shift in the way the existing work flow structure of design and construction works.

- An effective communication cycle must be established and maintained between major project participants as information exchange helps eliminate rework and consequently reduces time.

- Positive involvement of external as well as internal customers must be sought at the project early stages so that their requirements can be captured and implemented as input in the planning stage.

- Quality assurance techniques must be developed and implemented across the various work elements of the construction process.

- Innovations should be encouraged in areas of planning, contracting, design and construction.
- New approaches should be investigated to improve construction output.

Whereas all commentators are in agreement that the success or failure of reengineering hinges on the achievement of a cultural shift, there is less agreement on the detailed mechanism for bringing this about. Petrozzo and Stepper[18], for example disagree with Hammer and Champy 'that the leader of the reengineering project does not have to be involved in a near full-time capacity'. There are also differing views on the appropriate management style to be adopted. Petrozzo and Stepper[18] argue that 'the reengineering leader must *aggressively* communicate the importance of the reengineering effort, what it is going to do, and when it will be implemented' (our italics). To some, aggressive communication would seem to be a contradiction in terms and somewhat at variance with Coulson-Thomas's more spiritual view that 'organisations are living communities of people. They are sensitive organisms, reflecting, our dreams and fears. Knowledge workers are increasingly attracted to those networks whose values they share'[4].

Because of the philosophical and cultural differences which exist between, say European and North American corporations, it is not possible for us to give an explicit method of reengineering which will hold good for all situations. If we take the case in point of Petrozzo and Stepper versus Coulson-Thomas then some employees may respond positively to a hard-sell dictatorial approach which in other organisations would be a recipe for disaster. It is perhaps easier to say what the results of the reengineering process looks like rather than to attempt to describe a step by step progression through the implementation process. Even here however, Hammer and Champy[2] state that is not possible to give a single answer to 'what does a reengineering business process look like?' They do agree however that, not withstanding this caveat, there are a number of points of agreement which, by and large, typify the results of implementing reengineering. These are:

- several jobs are combined into one
- workers make decisions
- the steps in the process are performed in a natural order
- processes have multiple versions
- work is performed where it makes the most sense
- checks and controls are reduced
- reconciliation is minimised

- a case manager provides a single point of contact
- hybrid centralised/decentralised operations are prevalent

The explanation of the above set of characteristics is given in the context of North American business corporations and production processes which are taken from the factory floor.

In the construction industry context the following potential time and cost savings which would result from the application of reengineering have been identified by Mohamed and Tucker[9]. (This paper is of particular interest because it considers BPR at an industry level (in this case the construction industry) as opposed to the more common focus of BPR at the level of the firm. Readers are recommended to read this paper in full.)

Potential time saving

- Reducing the extent of variations and design modifications as a result of the systematic consideration of customer requirements in a clearly and accurately developed design brief.
- Producing less timely and more effective solutions whilst meeting client requirements by involving other engineering disciplines in the design phase.
- Achieving an agreed competitive overall construction time by negotiating better ways of construction with a contractor, who is selected on the basis of recognised past performance and financial stability.
- Improving the quality of design by incorporating the contractor's input in the
- design phase. This is conducive to smoother site operations and less construction delays.
- Compressing overall construction time by considering constraints imposed by downstream operations such as building approvals, material availability and site conditions during the design phase.
- Increasing the productivity of sub-contractors by re-organising small work packages into larger ones thus minimising delays caused by poor co-ordination and interference.
- Improving project performance by enhancing the working relationship between project participants through the adoption of team building, partnering and strategic alliance concepts. These concepts minimise the possibility of construction delays caused by conflict of interests.
- Increasing the efficiency of construction performance by reducing or eliminating inherent time waste in project material and information flows.

Potential cost savings

The majority of the above listed areas of potential time savings can lead, either directly or indirectly, to cost savings. In addition to the cost benefits attained due to the reduction of the overall construction time, other main potential areas of cost savings can be summarised as follows:

- Developing a design brief that accurately represents client requirements safeguards against additional costs due to design modifications, omissions and associated delays.
- Selecting a design option through the application of the value management concept implies avoiding more costly options for the same functional needs (clients get best value for investment).
- Applying the concept of concurrent engineering and considering the downstream phases of the constructed facility helps the client select a design option with less operating, maintenance and replacement costs (life cycle costs).
- Appointing a contractor based upon both past performance and financial stability reinforces to a large degree the process of controlling financial and operational risks.
- Implementing the concept of team-building between key project participants and partnering between the contractor and sub-contractors enhances working relationships and reduces the number of costly conflicts and claims.
- Adopting proper quality measures into the design and construction processes ensures minimising the amount of rework which the client ultimately pays for.
- Employing an efficient material management system, such as the Just-In-Time approach, save operating costs associated with material handling, storage, theft, damage, etc.
- Having project data communicated in a more timely and accurate manner reduces the possibility of extra costs resulting from decisions based upon poor or outdated information.

Pitfalls of reengineering

At the heart of reengineering lies the systems approach, with its holistic perspective. One corner stone of systems theory is that if one attempts to optimise the components of a system in isolation from one another then the inevitable outcome will be a sub optimal solution for the system as a whole. 'Undertaking self-contained BPR exercises at the level of the individual process can actually reduce the prospects of wider transformation. As a consequence of making an existing form of organisation more effective, the impetus and desire for an overall transformation may be reduced[4']

Because reengineering is very largely concerned with the breaking down of conventional subdivisions, for example, the reengineering case manager who has the power to leap departmental walls, then it is inevitable that the relationship of subsystems to the system as a whole will be a continuing preoccupation of reengineering practitioners. An instance of the suboptimal tendencies of BPR is where a BPR empowerment drive which encourages and trusts people to be flexible and catholic in how they work can be in conflict with a parallel BPR initiative seeking to define particular ways of approaching certain tasks. Or take a further instance where market testing can result in the carving up of an organisation into a collection of contractual agreements of varying time scales with the result that, in effect, there is no longer an organisational whole to reengineer or transform[16]. The natural reaction in the systems approach to preventing suboptimal solutions is simply to keep widening the system boundaries in order to be all-encompassing. Increasing the systems boundaries is not however always a practical, or even a conceptually desirable condition.

This is not to infer that there is an inherent conceptual flaw in the reengineering philosophy, but simply to point out that the self-induced challenges of reengineering are substantial, if not to say daunting. Most commentators would agree that it is a high risk and high gain occupation[11].

At an operational level Hammer and Champy list the following as mistakes to be avoided in the application of reengineering.

- Trying to fix a process instead of changing it
- Not focusing on business processes
- Ignoring everything except process redesign
- Neglecting peoples' values and beliefs
- Being willing to settle for minor results
- Quitting too early
- Allowing existing corporate cultures and management attitudes to prevent reengineering from getting started
- Trying to make reengineering happen from the bottom up
- Assigning someone who doesn't understand reengineering to lead the effort

- Skimping on resources devoted to reengineering
- Burying reengineering in the middle of the corporate agenda
- Dissipating energy across too many reengineering projects
- Attempting to reengineer when the CEO is two years away from retirement
- Failing to distinguish reengineering from other business improvement programs
- Concentrating exclusively on design.
- Trying to make reengineering happen without making anybody unhappy
- Pulling back when people resist making reengineering changes
- Dragging the effort out

Whilst the list is an informative set of instructions in its own right, it is also a useful insight into the pitfalls which can occur in trying to introduce reengineering into an organisation and also gives a very clear view of the traps which exist for the unwary.

A case study on the introduction of reengineering into a multinational electronic component company in a project named 'Smart Moves' gives the following lessons learned:

'Reengineering is not an easy or automatic activity. If it is to take place it cannot be done half-heartedly or in half measures. Although the methodology of selecting the process to be reengineered is not difficult to put in place, the execution of this is not an easy matter. Firstly, an appropriate process must be selected, capable of adding values and which can be clearly and concisely defined. The team must be empowered and an executive champion found. Additionally, getting a thorough understanding of the current process and representing this clearly and unambiguously can be tedious. Moreover developing a vision of the reengineered process through to the negotiating and executing the plan can be fraught with difficulty. A radical change may be identified by the team but this may be very difficult for management to commit to a high risk project, especially one which alters the status quo.'

As a counterpoint to these perceived difficulties the authors note that the reengineering process has serendipitous effects which are difficult to define but none the less exists. They conclude by remarking that 'Even if the radical changes required by BPR are not feasible, business process enrichment may enable existing processes to add value whenever possible[19].' A concluding remark, which seems to be somewhat rueful.

Information technology and reengineering

Just as reengineering is seen as a means of empowering modern business to affect radical change, so information technology (IT) is seen as the enabling mechanism to allow radical change to be affected. Many believe that reengineering is intrinsically linked to the application of IT[20]. Although 'BPR does not absolutely require the use of IT at all. However, one of the distinctive features of BPR is the way that IT is almost invariably used to 'informate' in Zuboff's terminology, to reconstruct the nature of work. It is a key feature of implementing BPR[21]'. There is a 'push-pull' synergistic relationship between reengineering and IT, with the power of IT extending the horizons of reengineering and the challenges of reengineering acting as a catalyst for new IT development.

IT can be extended to ICT (information and communication technologies)[22] and this is a useful broadening of the scope of IT, particularly when considering its relationship to reengineering. Several commentators[21,22] refer to the Venkatraman model[23] with respect to the relationship between IT /ICT developments and the business process.

The five stages of the Venkatraman model (Figure 3.2) have been described as follows[22]:

1. The first level of the model is termed 'localisation exploitation', and involves the employment of standard IT applications with minimal changes to the organisational structure of the firm. This level represents discrete 'Islands of Automation' within the business process.

2. The second level, 'internal integration' can be thought of as first building the internal electronic infrastructure which allows the integration of various tasks, processes and functions within an organisation and then uses this platform to integrate these intra-organisational processes. Venkatraman argues that this is an evolutionary step from 'localised exploitation'.

3. The third, and first of the revolutionary levels, business process redesign uses ICT as a lever for designing new business procedures rather than simply overlaying the technology on the existing organisational framework.

4. The fourth level, business network redesign involves the use of ICT to step beyond traditional intra-organisational boundaries to include clients, suppliers, and changes of the competitive environment by introducing a need for greater co-operation between trading partners.

5. The final level, business scope redefinition is the end point to the model which in Hazlehurst[22] suggests the question 'What role, if any, does ICT

have to play in influencing business scope and the logic of business relationships within the extended business network?'

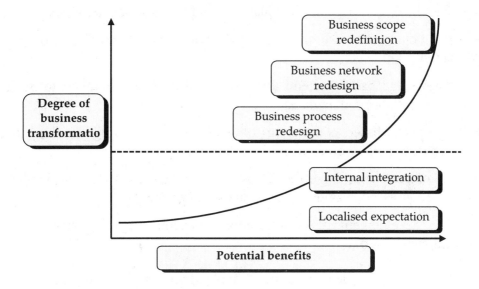

Figure 3.2 Five levels of itemised business

The Venkatraman model is useful in categorising levels of IT-enabled business particularly with respect to the threshold between evolutionary and revolutionary levels. Holtham[21], rightly, stresses the need to distinguish between the term 'process' as applied to business process and the term 'process' as it is applied in the field of computer science and software engineering. He makes the point that some IT advocates have used level 4 of the Venkatraman model as a justification for merely upgrading or replacing existing computer systems, which, of course, is at variance with the thrust of the Venkatraman model and with the philosophy of reengineering.

There are very few examples of the implementation of reengineering in the construction industry, but this not to say however that there are no examples to-hand of IT or ICT enabling technology which could be deployed in reengineering activities. Although the advancement in computer technology and widespread use of computers in the industry have opened a new frontier in construction applications, applications are still far from being satisfactory due to historical, financial and operational reasons. Not withstanding this caveat, Mohamed and Tucker[9] have identified the following IT technologies as enabling reengineering in the construction industry:

- CAD conferencing which allows project participants, e.g. the client, design team and contractors, working from different geographical locations and different hardware and CAD systems to work collaboratively.

Future developments in this area will link design offices and overseas construction sites[9, 24].

- The use of virtual reality for planning and construction activities.

- Relational data base shells for use in a range of building procurement activities.

- Knowledge based systems for use in building codes linked to Standard exchange for product data.

- Computer based simulation of work flow allowing leading to the application of automated and semi-automated robots.

- Automated cost engineering using modular, computerised knowledge bases accessed through a graphical user interface.

- The development of organisational decision support systems using a graphical interface.

The above examples of leading edge IT applications are focused, in the main on the planning and development stage. The following examples relate to the simulation of construction site operations[25].

- statistical process simulation of construction operations

- the use of graphical simulation to create a virtual site environment

- the use of simulation in the application of 'lean site' operations encouraging the use of resource minimisation techniques such as just-in-time.

Care has to be taken that because of the nature of the technology, IT has an seductiveness which may in fact detract from, rather than enable, the reengineering effort[26]. Whilst many companies have used IT in the furtherance of successful reengineering, 'Equally, there are a number of examples of organisations that used little or no new investment in IT to achieve their reengineering goals. In such cases the aim is to:

- Take IT off the critical path

- Make use of existing systems as far as possible

- Use flexible front ends to provide the appearance of seamless common user access to a range of underlying systems

- Only redevelop once the process has been redesigned and the new IT requirements, if any, are clear to all involved.'

This seems sensible advice in preventing the misuse, or wrongful application of IT in reengineering. One of the classical problems in computer applications, which has existed long before the introduction of the reengineering concept, is whether to use the computer to automate the existing manual system or whether to take the opportunity presented by a new computer technology, to re-think the existing process. The tendency has been to firstly automate the existing procedure, then at a later stage to recognise that an opportunity has been lost to really harness the power of the computer and then to backtrack to radically re-think the process. The golden rule as far as IT is concerned is only redevelop once the process has been redesigned, and the new IT requirements, if any, are clear to all involved. IT has a powerful role to play as an enabler, but not an instigator of reengineering.

Reengineering from a European perspective

As the reader will now have gathered, reengineering has much to do with cultural values. It is, by its very nature, an emotive concept. As we have stated, it has been our intention to try to steer a middle course in presenting the current reengineering movement without letting our own personal views intrude. We have included reengineering in this book because we believe that it is an important modern management movement and one worthy of serious consideration by the construction industry. However, it would be remiss of us if we were not to include some of the quite severe reservations which are held by some senior managers in Europe in terms of the impact and potential of reengineering.

The head of Siemens, Heirich von Pierer, writing on Business Process Reengineering has been quoted[21] to the effect that 'I don't feel completely comfortable with the radical thesis of Mr. Hammer. Our employees are not neutrons, but people. That's why dialogue is important.'

Holtham[21] goes on to express the view that BPR needs to be rooted in distinctive European managerial features, in terms of the 'acceptance of the humanistic and holistic stream in European thought, in contrast to the more mechanistic and fragmented US approach with the promotion of the concept of collaboration, between levels in the organisation, across organisations, between supplier and customer, and also across national boundaries. There would appear to be little variance between Holtham's vision of BPR in Europe and Hammer's description of the application of BPR in the States. Perhaps it is more a disagreement on the method of delivery than the message being delivered.' As Holtham concludes, ' The core elements of BPR have value beyond the evangelical North American approach, and BPR has value for Europe, if it is set in a European context'.

A case study of a process reengineering study in the Australian construction industry

There is no single answer to 'what does a reengineering business process look like?' Given that the very nature of the approach is dependent on creative thinking and the breaking down of existing structures and barriers, it is not possible to have a specific procedure or checklisting approach on how to implement reengineering. The following case study does however capture many of the features of reengineering through the initiation, planning and implementation stages.

(The text of this case study has been extracted by the authors from the final T40 report. We are indebted to the T40 contributing organisations for giving their permission to allow us to reproduce this material. For a more detailed discussion of the T40 project readers are referred to Ireland[27].)

Background

The T40 research project[27] was a process reengineering study in the Australian construction industry whose objective was the reduction of construction process time by 40%.

The participating organisations are shown in Table 3.3. This grouping represents three major contractors, two major services suppliers, two key consultants and CSIRO (the Australian national research and development organisation). The study was conducted in 1993/94 and reported in May 1994.

Table 3.3 Participating organisations

Fletcher Construction Aus.	CSIRO	BHP Steel
CSR	A W Edwards	Stuart Bros.
James Hardie Industries	Otis	Smith Jesses Payne Hunt
Taylor Thomson Whiting	Sly & Weigall	Motorala (USA) facilitators

Research method

The T40 team was lead by Professor Vernon Ireland of Fletcher Construction Australia, with each of the contributing companies havsing responsibility for an aspect of the process. Funding of over AU$300,000 was provided with the largest portion (AU$96,000) coming from the Australian Building Research Grants Committee.

The T40 group engaged Motorola (USA) to facilitate the analysis of the construction process as a process re-engineering exercise, drawing on Motorola's experience in process re-engineering their own manufacturing and project management operations. Motorola was also able to provide a non-construction industry perspective, which had no allegiance to the conventions of the construction industry.

The T40 study had three focal weeks of workshops which explored:

- Flow charting the 'As-is' process
- Redesigning the 'Should-be' process
- Developing aspects of the solution

All members of the team contributed to the final document. The document reflects the experience of team members, who represent most of the participants in the design and construction process. The intention was to produce a redesigned process and a series of new practices rather than simply complete present practices faster. Some of the proposals can be implemented with little disruption to current practices while some will require extensive discussion with other key parties in the industry before implementation.

Themes for the T40 project

The research findings included a number of key themes:

- Organisation of the T40 solutions team into a small group of up to 9 organisations, rather than a large group of general contractor plus specialists (50-100), with each T40 team member addressing the clients needs directly
- Reorganisation of work packages to eliminate multi-visits to the same construction location by the same specialist
- Maintenance of single point accountability for the client
- Financial incentives and penalties for the whole group of nine (including architect and structural engineer) to focus on action
- Business practices based on trust and fair dealing, thus eliminating the checkers, who are checking on other checkers on other organisations
- Getting it right first time with the elimination of rework.
- Elimination of traditional tendering, thus eliminating the time and cost of tendering as well as allowing the solutions team to directly answer the customer's needs.

- The solutions group/team sharing resources rather than duplicating functions (e.g. planning, supervision, employment of licensed trades and administration).
- Teaming between management and the workforce.
- Partnering with local government for approvals.

The T40 team recognised that changing the attitudes of key participants to achieve the innovations will not be easy. Many people in the industry will say that the goals are unachievable. However the team believes that such changes are necessary for the development of the Australian industry and it is feasible to achieve the goals over a period of time. The achievement of such goals will put Australia in a key position to take a strategic role in assisting the large industries in Asia.

Proposed T40 process

Agreed common goal between customer and delivery team

The T40 proposal is that all members of the solutions team will directly address the customer's needs, rather than have them filtered through the architect and filtered again through the general contractor. The whole team will directly focus on providing a solution which adds value to the customer's business.

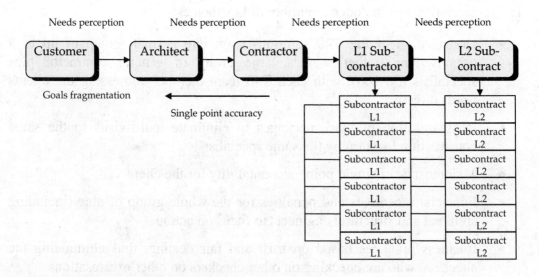

Figure 3.3 Traditional process hierarchy

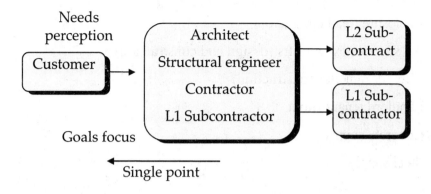

Figure 3.4 T40 solutions

Both the customer and the solutions team members need to state what they want from the process and the limitations to satisfying the goals of the other. To achieve such an approach the following changes to the status quo are needed:

- New methods of dependence and trust must be evolved, which will be enhanced by hunting for projects as a team together with financial incentives to support team play (shared success bonuses and penalties).

- New work patterns to reflect the dependence must be developed, for example the acceptance of a reallocation of traditional work packaging.

- Commitment must be obtained from the client and participant team at the highest corporate level.

Simplified Process

The proposed T40 project delivery system is a one-stage design and construction process with the whole solutions team involved from the point of determining the customer's needs. The key issue is that the customer is satisfied that value for money is being achieved and the project is delivered in the agreed reduced time.

It is a fundamental belief of the T40 research team that a construction team of 50 to 100 separate contractors is not as efficient in the long term as a group of 8 to 10 key players. Project managers with spans of control of 50-100 subcontractors and suppliers can only fight fires and have limited scope for innovations.

The T40 proposal is that, in a typical high rise office development, one contractor would be responsible for each of the following:

- Site establishment (lead contractor)
- Excavation (lead contractor)

- Structure (lead contractor)
- Enclosure (lead contractor)
- All services except lifts (design and construction)
- Lifts (design and construction)
- Landlord tenant fitout
- Concrete supply
- Steel supply

The architect and structural designer are also full members of the team, with ownership of goals and taking an appropriate share of risk and reward.

During the fitout phase there would be two services associates and two fitout associates. While this arrangement does introduce a three-layer hierarchy for the fitout and services contracts, the traditional reasons for there being no scope for control and performance incentives for these specialists can be overcome by separation in time and location.

It is essential that there be a clear definition of work packages between associates to eliminate interference and allow risks and rewards to be adequately apportioned.

Innovation that will occur within the package subcontracts will include:

- Development of innovative physical systems.
- Better integration of activities between the third layer of subcontractors.
- Re-engineering the sub processes to eliminate co-ordination problems.
- Analysis of the cost drivers in the process.
- Relation of cost drivers to value to the customer.
- More detailed planning, specification and integration of the activities of each construction worker.

The potential disadvantages of the arrangement could be problems in co-ordination and difficulties though a middle level organisation keeping the lead contractor from the subcontractors who are doing the work, and taking a margin for doing so. Ideally each of the services and fitout contractor should directly employ labour for their package of work. However, if they were not able to do so, because of fluctuations of workload, there must be incentives for the middle level contractors, in a three-level structure, to innovate and add value by their co-ordination.

Examples of construction process reengineering

(This is one of the three examples given in the T40 research report)

The three electricians

Under the current arrangements it is possible for three-electricians, employed by different subcontractors to be working on adjacent ladders in the suspended ceiling space. One is installing the lights, one installing the smoke detectors and one the air conditioning. They could each do 30 minutes work and then drive to the next project which by chance, happens to include electrical tasks, and so on and so on. A much more efficient approach would be for each electrician to do the full electrical work on each project.

An extension of this logic is illustrated in Tables 3.4 and 3.5 which itemise the traditional arrangements of subcontractors in the tenant space and the proposed T40 arrangements of package subcontractors in the tenant space:

Workforce empowerment

There are strong moves in many sectors of manufacturing, and in some sectors of construction, to redefine the traditional role of the foreman from that of planner and supervisor to that of coach, thus encouraging workers to be responsible for their own work, the planning of that work, the purchase and delivery of material for that work, and related activities. Specific aspects of workforce empowerment are workers who:

- are the eyes and ears of the foreman in identifying and correcting safety breached and QA issues;
- order their own equipment and materials;
- are represented on workplace committees responsible for resolving any potential disputes on site;
- are responsible for their own development to an agreed structure.

Teaming with the workforce

What is sought in a T40 environment is to achieve a fundamental cultural change which allows both the management and the general workforce to understand and respect each individual's or collective group's needs and subsequent goals.

The T40 solution embraces the concepts of teaming by:

Table 3.4 Traditional arrangements of subcontractors in the tenant space

Activity	Current Activities	By whom	Visit to area
1.	Air conditioning ducting	A	1
2.	Hydraulic rough-in	B	2
3.	Sprinkler pipes	B	2
4.	Steel stud partition frames to u/s slab	C	3
5.	Electric cable trays	D	4
6.	Smoke detectors rough-in	B	5
7.	Electrical rough-in	D	6
8.	Computer LAN rough-in	E	7
9.	Telephone rough-in	F	8
10.	Ceiling grids	G	9
11.	Tiles to sprinkler heads, AC regulators and lights	G	9
12.	AC fit-out	G	9
13.	Sprinkler fit-out	G	9
14.	Plaster board partitions	G	9
15.	Aluminium partition frames	H	10
16.	Glazing partitions	H	10
17.	Joinery	I	11
18.	Hang doors	I	11
19.	First coat of paint to partitions	J	12
20.	Skirting ducts	K	13
21.	Carpet and vinyl	L	14
22.	LAN fit-out	E	15
23.	Telephone fit-out	F	16
24.	Electrical fitout	D	17
25.	Smoke detector fit-out	B	18
26.	Ceiling tiles	G	19
27.	Second coat of paint to partitions	J	20
28.	Work stations	K	21
29.	Testing and commissioning power and lights	D	22
30.	Testing and commissioning air conditioning	A	23
31.	Testing and commissioning hydraulics & sprinklers	B	24
32.	Testing and commissioning smoke detectors	B	25

- Encouraging participating companies to develop a company enterprise agreement, either as a certified agreement or a formal commitment/ understanding, in relation to needs, goals and rewards.

- Establishing objectives and goals of not only the commercial stakeholders but participating workforce for the T40 project.

- Communicating this to all participants in the T40 project. They will often identify or relate to those that are common and can be identified in relation to their enterprise.

- Establishing methods of assessing the agreed objectives and goals that are being achieved.

- Encouraging workplace consultative committees and toolbox meetings to ensure continuing communication on what is or is not being achieved.

Codetermination

An extension of teaming is codetermination. A basic principle of codetermination is an absolute commitment to resolving industrial issues during the life of the project without resorting to lost time.

People innovation

An empowered group of people can be expected to make significant contributions to innovation on the project. Part of empowering people is to release these energies and thus encourage them to see the project as their own, so that they can contribute to the ideas bank on the project. Such arrangements could be the basis of a project enterprise agreement or an addendum to the company enterprise agreements.

Partnering with local government

The general theme of the T40 proposals is to partner with local government to satisfy both the needs of local government and the applicant. Some aspects of this proposal can be implemented immediately whereas some aspects will require longer term changes.

Longer term planning change

A longer term planning approval solution is for the building use and envelope to be specified as part of the zoning and, provided the proposed project meets the pre-approved guidelines, the planning approval could then be provided by administrative decision and action.

Table 3.5 Proposed T40 arrangements of package subcontractors in tenant space

Activity	T40 proposed activity	By whom	Visit to area
1.	AC ducting Hydraulic and sprinkler rough-in Smoke detectors Cable trays Telephone rough-in Electrical rough-in	A	1
2.	Computer LAN rough-in	B	2
3.	Ceiling grid Tiles to sprinkler heads, AC registers, lights Stud partition Plasterboard Aluminium glazing frames Glazing Joinery Hang doors First paint Skirting ducts	C	3
4.	LAN fit-out	B	4
5.	Mechanical and electrical fit-out	A	5
6.	Telephone fit-out	D	6
7.	Ceiling tiles, final coat to partitions	C	7
8.	Install work stations	E	8
9.	Testing and commissioning services	A	9

Tendering

The central assumption of the T40 process is that the full solutions team, including the general contractor and specialist contractors, are fully engaged in addressing the customer needs from the outset. This significant process requirement is quite different to lump sum tendering and even the current form of document and construct, of developing a basic design by external consultants and then calling tenders based on this outline design and performance requirements. Both of these current processes exclude the solutions team from directly addressing the customer's need.

The proposed process eliminates the waste of current tendering, with success rates for contractor of approximately one in ten. What is needed is a system that selects the construction contractor on the basis of:

- Past record on time performance by reference to an industry time database.
- Agreed cost based on an adequately developed design.
- Third-party endorsement of cost as being close to the best price.
- Agree time which is significantly better than the industry average, for a building of the type (e.g. 40% better time performance, which corresponds to a 25% reduction in costs).
- Public accountability.

These points are illustrated in Table 3.6.

Table 3.6 Differences between tradition, document and construct and the T40 solution

	Fully document lump sum	Document and construct	T40 Process
Contractor involvement in determining customer needs	No	No	Yes
Subcontracting involvement in determining customer needs	No	No	Yes
Single point accountability	Yes	Yes	Yes
Number of subcontractors and suppliers	50-100	50-100	Tiered with 8-10 in first layer
Incentives and penalties for subcontractors	No	No	Yes
Identification of subcontractors with owner	No	No	Yes
Clear stages of project approvals	Sometimes	Sometimes	Yes
Real time design	No	Sometimes	Yes
Variations in process	Usually many	Medium to few	Nil
Documentation process	Multi-origins	Some co-ordination	Staged, packaged & complete
Project communications	Phone & fax	Phone & fax	Direct
Use of specialist resources	Separate	Separate	Shared
Supervision	Separate	Separate	Shared
Automatic invoicing	Seldom	Seldom	Yes
Tendering	Costs of 6-8% & time of up to 3 months	Costs of 3% and time of up to 3 months	Virtually zero costs
Local authority approvals	6-12 months	6-12 months	0 months
Predictable outcome	No	Better	Yes
Overall work to complete	W	90%W	75%W
Time outcome	T	90%T	60%T

Conclusion

In this chapter, we have traced the origins of reengineering, we have examined its goals and detailed the tactics which have been adopted by successful practitioners of reengineering in the business community at large. We have also included a detailed description of the T40 project, which as a construction industry research study in process reengineering, exemplifies the key characteristics of the reengineering approach.

We have contrasted the directness of the North American approach to reengineering with the more restrained European view. Our own view of reengineering is one of cautious optimism. At present the very few examples of the successful application of reengineering in the construction industry prevent us from being more sanguine about its future. Having said that, we do note however that there is a ground swell in the industry towards the adoption of reengineering principles with the Commonwealth of Australia's Science and Industry Research Organisation (CSIRO), Division of Building, Construction and Engineering embarking on a three year research program with the aim of offering solutions to the reengineering of the construction process. To quote:

'CSIRO is committed to construction process re-engineering, recognising that it can significantly strengthen the Australian construction industry against overseas competition, both in the home and overseas market'.

Reengineering has the power to change the structure and culture of the construction industry. Time will tell whether or not the industry will rise to the challenges posed by the reengineering approach.

---

[1] Love P. and Mohamed S. (1995) Construction Process Reengineering. *Building Economist*, Dec. 8-11.

[2] Hammer M. and Champy J. (1993) *Reengineering the corporation: A manifesto for business revolution*. Nicholas Brealey, London.

[3] Morris D. and Brandon J. (1993) *Re-engineering your business*. McGraw-Hill, New York.

[4] Coulson-Thomas CJ, (editor) (1994) *Business process reengineering: Myth and reality*. Kogan Page, London.

[5] Hammer M. (1990) Re-engineering work: Don't automate, obliterate. *Harvard Business Review* July / August, 104-112.

[6] Rigby D. (1993) The secret history of process engineering. *Planning Review* March/April, 24-27.

[7] Whitney J.O. (1987) *Taking charge: Management Guide to Troubled Companies and Turnarounds*. Dow Jones-Irwin, Illinois.

[8] Greengard S. (1993) Re-engineering: Out of the rubble. *Personnel Journal*. December.

[9] Mohamed S. and Tucker S.N. (1996) Construction process engineering: potential for time and cost savings. *International Journal of Project management* (special issue on Business Process Reengineering).

[10] Construction Industry Development Agency. (1994) *Two steps forward and one step back: management practices in the Australian construction industry*: CIDA, Commonwealth of Australia Publication, Sydney NSW Australia, February.

[11] MacDonald J. (1995) *Understanding business process reengineering.* Hodder & Stoughton, London.

[12] Obeng E. and Crainer S. (1996) *Making re-engineering happen: Whatís wrong with the organisation anyway?* Pitman Publishing, London.

[13] DuBrin A.J. (1996) *Reengineering survival guide: Managing and succeeding in the changing workplace.* Thomson Executive Press, Ohio.

[14] Armstrong J.S. (1985) *Long-range forecasting: from crystal ball to computer.* 2nd rev. ed., Wiley, New York.

[15] COBRA (1994) *Business restructuring and teleworking: Issues, considerations and approaches.* (Methodology Manual for the Commission of the European Communities), London.

[16] Coulson-Thomas C.J. (1994) *Implementing re-engineering.* In: *Business process reengineering: Myth and reality* editor Coulson-Thomas. pp. 105-126, Kogan Page, London.

[17] Mohamed S. and Yates G. (1995) Re-engineering approach to construction: a case study, *Fifth East-Asia Pacific conference on structural engineering and construction - Building for the 21st century.* 25-27 July, Vol. 1, 775-780, Gold Coast, Queensland.

[18] Petrozzo D.P. and Stepper J.C. (1994) *Successful reengineering.* Van Nostrand Reinhold, New York.

[19] Fitzgerald B. and Murphy C. (1994) The practical application of a methodology for business process re-engineering. In: *Business process reengineering: Myth and reality* (editor Coulson-Thomas). pp. 166-173, Kogan Page, London.

[20] Betts M. and Wood-Harper T. (1994) Re-engineering construction: a new management research agenda. *Construction Management and Economics*. **12,** 551-556.

[21] Holtham C. (1994) Business process re-engineering: contrasting what it is with what it is not. In: *Business process reengineering: Myth and reality* (editor Coulson-Thomas) pp. 166-173, Kogan Page, London.

[22] Hazlehurst G. (1995) Re-engineering the processes of construction engineering and design: islands of automation. *COBRA 95 RICS Construction and building research conference*. 8-9 Sept., Vol. 2, 207-215, Heriot-Watt University, Edinburgh.

[23] Venkatraman N. (1991) IT-induced business reconfiguration . In: *The corporation of the 1990s* (editor Scott Morton), pp. 122-158, Oxford University Press, New York.

[24] Newton P.W. and Sharpe R., (1994) Teleconstruction: an emerging opportunity. *National construction and management conference;* Feb. 447-459, Sydney.

[25] Mohamed S. Simulation of construction site operations: a reengineering perspective. submitted to *Engineering, Construction and Architectural Management Journal*

[26] Talwar R. (1994) *Re-engineering: a wonder drug for the 90s?* in: *Business process reengineering: Myth and reality* (editor Coulson-Thomas). pp. 40-59, Kogan Page, London.

[27] Ireland V. (1994) *T40 process re-engineering in construction.* Research Report, Fletcher Construction Australia Ltd., May.

4 Partnering

Introduction

'Partnering' is difficult to define. It means many things to many people. Partnering has to do with human relationships, with stakeholders interests, with the balance of power. In other words partnering has to do with human interaction and as an inevitable consequence of this, it is a complex subject which is difficult to pin down and analyse. Partnering is more than simply formalising old fashioned values, or a nostalgic return to the good old days when a 'gentleman's word was his bond', (although moral responsibility and fair dealing is an essential underpinning of any partnership[1,2]). It is more than a building procurement technique (although building procurement techniques can be used to operationalise good practice, bring about cultural change and thus create a more cohesive team[3]). The use of partnering in the construction industry has many advocates and many claims of success. The titles of journal articles on partnering positively exude confidence and self assurance. Titles such as 'Partnering means making friends not foes[4]', 'Partnering pays off[5]', 'Partnering makes sense[6]' and more forcefully 'Partnering - the only approach for the 1990s[7]', abound in the professional journals.

In this chapter we will try to cut away the hype surrounding partnering, by tracing its origins, describing how it has been adopted by the construction industry and then describing the benefits and also the risk and pitfalls associated with the implementation of partnering.

The origins of partnering

The origins of partnering, as a construction management concept, are relatively recent, dating from the mid 1980s[8]. This not to say that partnering did not exist prior to that period and indeed many would subscribe to the view that 'Partnering between contractors and private clients is as old as construction itself'[9]. It has also been claimed that, in the UK, companies such as Bovis have developed a culture and tradition of non-adversarial relationships with particular clients since the 1930s[3].

For the purposes of this chapter we will concentrate on the period from the mid 1980s onward when the term 'partnering' was given quite explicit connotations. In effect we will focus on *formal* partnering, where there is evidence of an explicit arrangement between the parties. This is not to dispute the existence and importance of *informal* partnering (or as it has been described 'partnering without partnering'[10]). However for the time being we will discount informal partnering from our considerations.

According to the National Economic Development Office (NEDC) report 'Partnering: contracting without conflict'[11], true partnerships in the formal sense only became established in the mid 1980's, the first being that between Shell/ Parsons/Sip in 1984. The most frequently cited partnering arrangement of the 1980s is the Du Pont/ Fluor Daniel relationship for the Cape Fear Plant project. The partnering agreement between Du Pont and Fluor Daniel was made in 1986[11] and was a formalisation of a relationship which had existed since 1975. Other notable partnering relationships during this era were Union Carbide/Bechtel; Proctor & Gamble/ Kellog; and Shell Oil/ Parsons.

Partnering in a construction industry context

Most commentators attribute the emergence of partnering as a force in the construction industry in the late 1980s, to the work of Construction Industry Institute of the United States(CII) and the adoption of partnering by the US Army Corps of Engineers (mainly through the efforts of Charles Cowan)[1, 12, 13]. In the present era, extensive examples of partnering can be found in the Unites States with the movement gaining momentum in New Zealand, Australia and the UK. In the latter two countries this gain in momentum is partly as a result of prompting by the Gyles Royal Commission into Productivity in the Building Industry in New South Wales[13] and the Latham Report in the UK. Latham in his foreword to 'Trusting the team: the best practice guide to partnering in construction'[14] states that 'partnering can change attitudes and improve the performance of the UK construction industry. I hope that the industry and its clients will now use the report to embark upon partnering'.

In Australia the Gyles Royal Commission went further than this and carried out a pilot study on partnering as a means of encouraging a cultural shift in the New South Wales construction industry[13]. Partnering is one of the approved criteria in the New South Wales Department of Public Works and Services (DPWS) contractor accreditation scheme[15] demonstrating the importance which some government clients place on the introduction of partnering.

The extent of the adoption of partnering by the construction industry at large, is still difficult to quantify. There are however numerous examples from the US of successful partnering. A 1994 study[16] of 2,400 attorneys, design professionals and contractors rated project partnering and mediation as top of the list of alternative dispute resolution methods. The 1994 annual meeting of the Construction Industry Institute in Boston reported that 'in terms of programs such as safety, constructability, total quality management and both long-term and short-term partnering that long-term partnering offered the most impressive savings. CII's strategic alliance task force reported that 196 projects using long-term partnering saved on average of 15% of total installed cost. On project-specific partnering, CII's team-building task force recorded an average of 7% savings on five very large projects'[17]. These statistics indicate the general level of the acceptance of partnering in the United States.

In Australia partnering is relatively common place to the extent that the Master Builders Association runs an annual competition which attracts a range of entries of good examples of partnered projects. Hellard[1] is somewhat sceptical of the extent of the uptake of partnering in the UK. 'A full partnering project embracing client, design team, general contractor, and sub-contractors with a formal project charter and formal training and commitment sessions does not yet seem to have emerged'. However several case studies of UK partnering are included in the recently published 'Trusting the team: the best practice guide to partnering in construction'[14].

The goals of partnering

We said at the onset of this chapter that partnering is a difficult phenomenon to isolate and define. However defining the *goals* of partnering, as opposed to defining the *nature* of partnering is simpler.

There are a number of definitions in circulation on the goals of partnering. Some of these are very broad, for example 'partnering is a process for improving relationships among those involved in a construction project to the benefit of all'[18]. Others are much more detailed but share the same philosophy, for example:

'[Partnering] is not a contract but a recognition that every contract includes a covenant of good faith. Partnering attempts to establish working relationships among stakeholders through a mutually developed formal strategy of commitment and communication. It attempts to create an environment where trust and teamwork prevent disputes, foster a co-operative bond to everyone's benefit and facilitate the completion of a successful project'[19].

All commentators stress the achievement of trust and co-operation as the essential goal of partnering. Typically this is described as 'a long term contractual commitment between two or more organisations based on a spirit of trust and co-operation. The idea is to allow each participant to make the most of his resources and continually improve performance'[20].

Cowan, one of the principal architects of the modern partnering movement stresses that 'Partnering is more than a set of goals and procedures; it is a state a mind, a philosophy. Partnering represents a commitment of respect, trust, co-operation, and excellence for all stakeholders in both partners' organisations[12].'

Categories of partnering

There are two different categories of partnering and within these categories there are a variety of types. The two categories of partnering are: strategic partnering and project partnering. (Strategic partnering is sometimes referred to as 'multi-project partnering' or less frequently as 'second-level partnering' and project partnering as 'single project partnering' or 'first-level partnering'.) 'Strategic partnering takes place when two or more firms use partnering on a long term basis to undertake more than one construction project'[14]. Project partnering is the converse of this, and occurs when two or more firms come together in a partnering arrangement for a *single* project. In the United States 90% of all partnering is project partnering[14]. Because partnering has to do with long-term relationships, it follows that more gains and benefits are likely to be achieved from the longer term strategic partnering as opposed to the shorter duration project partnering arrangements. However project partnering can be a stepping stone to strategic partnering.

[Note: For the purposes of this text we use the term 'project' to mean a construction industry project of a 'normal' timescale. Previously we cited the Du Pont/Fluor Daniel, Cape Fear Plant project as an early partnering agreement. Although this was a single project, it was of such massive dimensions, both in terms of physical and timescale, that this would most properly be categorised as strategic rather than project partnering. This however is the exception which proves the rule. In most cases the differentiation between 'project' and 'strategic' is relatively straight forward.

Project partnering

The participants

The first issue to be addressed in any form of partnering, is who should participate in the partnering arrangement? This can immediately give rise to a 'chicken and egg' syndrome. If the partnering concept is introduced at the outset of the project then this will have a bearing on the procurement method adopted which in turn will have a bearing on the composition of the project participants. In order to gain the maximum benefits of partnering, the general trend is to use a design and build (design and construct in Australia) form of procurement. The advantage of this approach is that it allows all key stakeholders i.e. the client and the design and build contractor (whose organisation includes the design team), to be involved in a partnership arrangement from the onset of the project.

Partnering is not however the exclusive domain of design and build and can be used in a traditionally procured project where the lowest bid contractor is brought on board at the tender acceptance stage. In this situation the contractor is excluded from the design stages, thus during the pre-construction phase the partnering arrangements can only take place between the client and the design consultants, the contractor being brought into the partnering relationship at the downstream stage of tender acceptance. This is not the ideal environment for partnering and has been likened to bringing a new player into a bottom of the league football team half way through the season and then expecting them to become league champions[10].

The selection of the members of the partnering team is clearly of paramount importance in a concept which is all about trust and mutual support. The partnering team must be capable of carrying out its responsibilities throughout the project. 'If any one of the team members is not capable of carrying out its duties, any attempt at partnering will fail'[10]. However the task of selecting partners is not an easy one. As Cowan remarks 'Ideally, you want to select contractors or owners who have established a successful track record of partnering on previous contracts. By interviewing contractors an owner can discern interest and/ or expertise.' At first glance this statement may seem to be a tautology. However it is possible for a contractor to progress from being an interested party to a partnering participant. This progression is described in the New South Wales Contractor Accreditation Scheme[15] which identifies 5 categories of partnering involvement ranging from low to high. These categories are as follows (Table 4.1):

Commitment

The decision to partner is a very significant one because of the reliance which one partner must inevitably place on the other/s. It follows that a decision to partner should be taken at the highest possible level in an organisation. 'It is inconceivable that....partnering could be carried out without commitment at the very highest levels within both organisations'[11].

Although great stress is made by all commentators on the need for high level commitment there is also a need for 'internal partnering'to take place within an organisation prior to a commitment by senior management to enter into a partnering agreement. 'Internal partnering means preparing your organisation and reviewing internal procedures and documentation. It also involves educating and informing staff on partnering. The result of this process should be a united organisation that is prepared to work closely with another organisation'[14].

Table 4.1 Levels of partnering characteristics[15]

Level	Stage	Characteristics
Low	Recognition stage	Aware of benefits of partnering
Average	Development stage	Developing partnering policy
Above average	Establishment stage	Committed to use of partnering in all projects
Good	Continuous improvement stage	Documented evidence of ongoing improvement in project delivery through partnering
High	Best practice stage	Record of long term relationships with consultants, suppliers and sub contractors

The partnering process

It is inevitable that the partnering process will have many forms and many variations. In this section we are considering single project partnering, the processes for which will be different to strategic partnering which we will consider later. The process which is adopted for project partnering will depend on the circumstances of each situation. If the project participants have previously been involved in similar partnering arrangements then many of the bridges will already have been made, whereas if this is a first-off occasion for the participants then the process will have to be developed with particular care and sensitivity. For the purposes of this section we will assume that we are dealing with partnering with first time partners.

Fundamentally the partnering process is about team building which is why the function of internal partnering is so important in achieving a successful outcome. The processes of partnering must be seen as a means to an end and must not be seen as an end in itself. Katzenbach and Smith state that 'a demanding performance challenge tends to create a team. The hunger for performance is far more important to team success than team building exercises, special incentives, or team leaders with ideal profiles'[21]. It is essential that the team is enthused by the challenge of partnering, and the generation and maintenance of this enthusiasm is one of the primary functions of the partnering process[1].

In general the partnering process falls into three phases. The pre-project stage, the implementation stage and the completion or feedback stage.

Pre-project stage

The pre-project stage begins with the decision of whether or not to partner. In some cases a readiness to use partnering will be a condition of pre-qualification for government tenders[15]. Whether or not the initiative to partner comes from the client, design team or contractor, it is a potentially high risk decision and as such cannot be taken lightly. It is not an automatic assumption that partnering is always the best approach. For some parties and in some situations partnering should be best avoided. Ideally there should be a synergistic relationship between the parties with each party bringing complementary strengths to the partnering table. The worst case scenario is where an inherent weakness in one party spreads throughout the team, in this situation partnering is likely to exacerbate rather than stabilise the situation.

Assuming that the decision to partnering is taken, how then is the process of selection be initiated? There is no clear answer to this question. In some situations, such as the New South Wales Department of Public Works and Services (DPWS) in Australia, partnering may be a direct requirement of prequalification to tender for government projects. The partnering connections must however be made by the individual parties, albeit with the encouragement of DPWS. In other instances the initiative to partner may come from an individual within a company. Award schemes such as the Australian Master Builders partnering awards help to sow the seed, as does publications such as 'Trusting the team: the best practice guide to partnering in construction'[14]. Given that partnering has to do with trust and mutual co-operation, the process of making the initial overtures between potential partners is never going to be prescriptive and to some extent it will always be shrouded in mystery and cloaked with commercial confidentiality. The following overture from a client to a design and build contractor gives a flavour of how a partnering arrangement could be initiated:

'In order to accomplish this contract most effectively, the Owner proposes to form a cohesive partnership with the Contractor and its subcontractors. This partnership would strive to draw on the strengths of each organisation in an effort to achieve a quality product done right the first time, within a budget and on schedule. This partnership would be bilateral in make-up and participation will be totally voluntary. Any cost associated with effectuating this partnership will be agreed to by both parties and will be shared equally with no change in the contract price[22].'

As can be seen this is voluntary agreement which is aimed at creating a culture of mutual trust as opposed to an adversarial climate. (The reader may feel at this stage that, given that, the conventional construction contracts are, of their very nature, adversarial in approach that this presents something of a dichotomy between the goals of partnering and the normal contractual agreements. This is a difficult issue and will be dealt with later in the chapter under contractual and legal issues. For the moment we will leave these qualms aside.)

Having achieved an understanding to partner between top management, the progression of the partnering process takes a more predictable path. Figure 4.1 is a stylised flow chart of the partnering process.

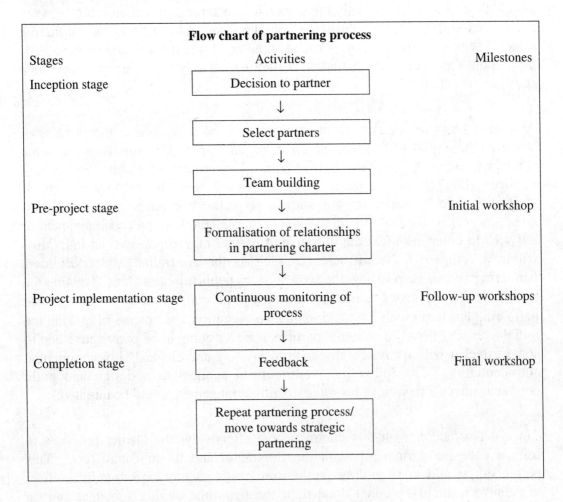

Figure 4.1 Stylised flow chart of the partnering process

Initial partnering workshop

Once the decision to partner has been made, the stage is set to bring together the key middle managers who will be involved in the project on a day to day basis. In a design and build procurement approach this group will comprise a range of stakeholders from the client body, the design team and the contractor. In addition there may also be legal and financial advisers and on some occasions even representatives from statutory planning authorities and building control. In the interests of group dynamics the size of the group should not be overly large and should be restricted to a maximum of about 25[14]. The team is normally introduced to one another through an initial partnering workshop (the detailed mechanisms applied at this workshop will vary according to specific circumstance and also according to the culture of different countries.) The purpose of this workshop is to agree the ground rules for the way ahead and should be held as soon as possible after the contract has been awarded. There seems to be general agreement that this workshop should be held on neutral territory, preferablely at a 'retreat', away from all other distractions and would normally be of one, two to three days duration depending on the nature of the project and familiarity of the participants with partnering. It is highly recommended that the workshop sessions are managed, for at least part of the time by a facilitator with experience of partnering.

The goals of 'the workshop are: to open communications, develop a team spirit, establish partnering goals, develop a plan to achieve them, and gain commitment to the plan'[22]. The general pattern of the workshops is to begin with team building exercises, where participants are encouraged to explore mutual interests of a personal nature. This may be done in quite a low key way, or may be more formalised by using devices such as the Myers-Briggs Type indicator as a means of personality profiling. The team then moves on to develop a conceptual framework or common mental picture of what partnering is about and how this will influence their own participation in the project. The team then goes on to consider and agree mutual objectives and mutual interests on the lines of a SWOT analysis. (SWOT analysis means identifying Strengths, Weaknesses, Opportunities and Threats.) Mutual objectives such as time, cost and quality will feature large on the agenda with other issues such as safety, constructability, electronic data interchange (EDI) also being explored. The mechanism for arriving at these objectives is usually to form action teams from within the participants to deal with the many issues which will emerge during the course of the workshop. These action teams may comprise the counterparts of the disciplines of each of the partners, for example legal advisers, or may contain a mix of disciplines such as architects and engineers. In any event the action teams will be tasked with bringing solutions to the group as a whole on issues such as, deriving performance indicators, dispute resolution or minimisation and safety issues. The action teams assist the group bonding process and reduce the partnering issues to manageable dimensions. In many instances the members of the action teams will remain in close communication throughout the duration of the project.

The first workshop session culminates in agreement of a partnering charter and in agreeing a partnering implementation plan. Agreement should also be reached on the timing of follow-up workshops which are essential in maintaining the will to succeed in the partnering arrangements. The action teams will be tasked with the development of their assigned topics for the follow up work shop, which would typically be held at three monthly intervals over the duration of the project.

A partnering charter (or project charter) will always be signed by the participants. (The signing of the charter by all the participants is seen as a very symbolic act and is a tangible way of swearing an oath of allegiance to one another and to the recognition that individual interests are subordinate to those of the project.) The signatories to the charter (in a design and build scenario) could come from the following organisations: client, contractor, design team, subcontractors, manufacturers, suppliers, consultants (such as legal advisers and planning experts).

The actual charter may be quite general or quite detailed depending on the preferences of the partnering team and the nature of the project. Given that copies of the charter will be posted on the walls of all the participants and be displayed in a variety of locations from the client's office to site workshops and site offices, then a single A4 sized sheet is the limiting factor. An example of a partnering charter is shown in Figure 4.2:

To summarise, irrespective of the mechanism used to forge a partnering agreement, 'before the project starts, participants need to establish a common charter for the project, agree on performance criteria and how they will be measured, develop mechanisms for resolving conflict, and establish general guidelines for working together. Considerable time and energy must be invested up-front to establish the foundation for a working relationship that can tolerate conflicts and avoid costly misunderstandings'[12].

Implementation stage

At the conclusion of the initial partnering workshop an implementation plan will have been established which will include a timetable for a series of follow-up workshops. The frequency and the duration of these workshops will depend on individual circumstances. The initial workshop is geared towards changing a traditional adversarial culture to a team spirited win-win environment. This quantum leap in human relations needs constant nurturing (particularly with newcomers to partnering). Follow-up workshop sessions are essential in reinforcing the partnering culture. The follow-up workshop sessions usually follow a similar pattern to the initial workshop. If an external facilitator has been used for the initial partnering workshop then it makes sense to use the same facilitator for the follow-up sessions and, as with the initial workshop, it is recommended that the follow-up sessions are held on neutral territory away from distractions.

Partnering Charter

We the Partners of XXXXX Project are committed to working together in a trusting and sharing environment, and are dedicated to achieving excellence for the benefit of all the stakeholders involved in this project.

The goals of this project are to:

Complete the project on time

Complete the project within budget

Achieve a reasonable profit margin

Have zero lost time due to accidents

Produce a high quality building with zero defects

Our team mission statement is that:

We will work in an environment of open door communications

committed to achieving excellence in the XXXXX project

Signed by:

_____   _____
_____   _____
_____   _____
_____   _____

Figure 4.2 Partnering Charter

It is important that the parties to the partnering charter agree to evaluate the team's project performance against an agreed set of criteria. Cowan et al[12] suggests a simple but formal scaled questionnaire which, in addition to criteria such as cost control, safety record and time scheduling, also includes criteria which are explicit to the partnering relationships such as teamwork and problem solving. The Royal Commission into Productivity in the Building Industry in New South Wales, which introduced the partnering concept to Australia, used a 10 point partnering effectiveness monitor in its pilot study in 1992.

A typical partnering evaluation summary, or partnering effectiveness monitor, is shown in Figure 4.3:

Partnering evaluation summary								
No.	Objective	Last period			This period			
		Weight	Rating	Score	Weight	Rating	Score	Comments
1	Quality achieved	15	4	60	15	4	60	
2	Cost controlled	15	4	60	15	5	75	
3	Time performance	15	3	45	15	4	60	
4	Teamwork	10	4	40	10	4	40	
5	Safety	10	3	30	10	3	30	
6	Avoidance of industrial disputes	10	4	40	10	2	20	
7	Avoidance of litigation	5	4	40	5	4	40	
8	Satisfactory cash flow	5	5	25	5	4	20	
9	Environmental impact contained	5	3	15	5	3	15	
10	High morale and job satisfaction	5	4	20	5	5	25	
Overall satisfaction of partnership stakeholders		100	1 to 5	375	100	1 to 5	385	

Figure 4.3 Partnership evaluation summary

For projects of more than one year duration, formal written partnering evaluations might take place every 1-3 months with follow-up team meetings every 3-6 months. These formal evaluations will be a key item on the agenda of the follow-up workshops.

In addition to the scheduled follow-up partnering workshops, informal partnering sessions will also take place. Although these informal session are important in terms of maintaining good relationships between the partners, it is equally important that if and when new members join the team, that they are inducted into the partnering process via a formal workshop session rather than through an informal session. This is because there is always a danger that the ethos of partnering will become diluted (particularly with participants who are new to partnering) in the dissemination process.

Partnering pitfalls during the implementation stage

Project partnering is a high-risk, high-gain approach. Naturally many of the problems which can be identified during the implementation stage of the partnering process also apply to non-partnered conventionally procured projects. However although there is a greater potential for stakeholders (those involved in the partnering process) to gain from partnering compared to the conventional process, conversely, there is also a greater risk of loss to the stakeholders if things go wrong.

Kuball[10] lists six criteria as being essential to successful partnering. It could be argued that these criteria are equally essential to the success of a non-partnered process. However if these criteria, which are summarised below, are not in place then the partnered process is at greater risk.

- There must be no weak links in the team members. They must all be capable of performing their tasks. (Bennett and Jayes[14] go as far to suggest that team members be allowed two 'mistakes' after which consideration should be given to their being replaced.)

- Owners must be properly represented and active in the project.

- The composition of the partnering team should, wherever possible, remain stable for the duration of the project.

- Partnering should start at the design stage.

- There is a need to appoint a team leader or 'champion' to ensure that partnering principles do not slip out of focus.

- The involvement of major sub-contractors and manufacturers is crucial as is the integration of trade contractors.

These six criteria underscore the vulnerability of partnering relationships and highlight the challenges involved in implementing a successful partnering program. "The fragility of inter-organisational alliances stems from a set of common 'dealbusters' - vulnerabilities that threaten the relationship. Partnerships are dynamic entities, even more so than single corporations, because of the complexities of the interests in forming them. A partnership evolves; its parameters are never completely clear at first, nor do partners want to commit fully until trust has been established. And trust takes time to develop. It is only as events unfold that partners become aware of all the ramifications and implications of their involvement.[23]"

There is of course no concrete evidence that partnering will always benefit the project. Although on the one hand partnering can be viewed as "saving money and making work more enjoyable" there are also disadvantages[14]. Cost is one example. The direct costs of partnering are the cost of the workshop sessions and also the cost of the partnering facilitator, hiring of venues, transport etc. The cost of the participants' time in attending workshops etc. is not an additional direct cost of partnering but is a redistribution of costs which would normally occur, or at least should occur in the traditional procurement process. In addition other potential disadvantages and indirect costs can be identified as follows[14]:

- **Stale ideas** - due to the lack of stimulation which can occur when the same partners are in a stable relationships and the stimulus of new players is missing.

- **The cost of consultants may be higher** - the partnering process produces more alternative solutions. As a result design consultants have a greater work load which should be reflected in higher fees.

- **Reduced career prospects** - staff involved in partnering may see this as an additional burden in terms of work load, which will not count towards their career advancement.

- **Loss of confidentiality** - there can be problems associated with this unless proper safeguards are taken.

- **Investment risk** - Investment in joint development for the project, for example in developing an EDI (electronic data interchange) system, may be risky if the single project partnership does not extend into future contracts

- **Dependency risk** - there is a risk that the partners become too dependent on each other. They should therefore spread their partnering agreements over a range of partners. On occasion however short term financial aid may need to be given to a single source supplier in order to ensure the supplier's continued existence.

- **Corruption** - the one-off relationship of non partnered projects minimises the possibility and opportunity for corruption. There is an additional obligation on partnerships which may become long term in nature to ensure that safeguards and checks against corruption do not become lax.

The underlying cautionary note which links all of these potential disadvantages of partnering is that, to be successful, partnering needs an equal level of commitment from all the partners. 'Unevenness of commitment often develops from the basic differences between organisations. If for example, a small contractor is entering into a public works or defence contract managed by a large government agency, the contractor may feel that it is not possible to devote the time and staff to partnering that will make the contractor an equal partner. Every effort is needed by all involved parties to balance the commitment on both sides[22]."

If this commitment is given many of the pitfalls of implementing partnering will be avoided.

Limits to partnering

There is general agreement amongst commentators that participants in partnering should not put all of their eggs in the one basket, and should limit the extent of their partnering commitments. As early as 1981 the UK National Economic Development Office[11] recommended that 'In general it was considered that no single partnering arrangement should utilise more than 30 per cent of the resources of the office in which it was located, and that a contractor's total commitment to partnering should not utilise more than 50 per cent of total technical and managerial resources.'

Completion stage

The completion stage of the partnering process is particularly important, and is one of the particular strengths which the partnering process has in comparison to the traditional project team approach. In this latter type of arrangements there is rarely a formal procedure that attempts to encapsulate the knowledge base which has accumulated as a consequence of the project.

The completion stage of the project partnering process would normally be centred round a final workshop session. Usually this final workshop will incorporate a lunch or dinner at which senior management will take the opportunity to make mention of specific contributions and highlight the characteristics of the completed project. Although this social event is an occasion for emotional release and for self-congratulation, (particularly if the project has gone well) , the final workshop is not simply to do with social niceties. It is an opportunity to undertake some serious work in order to learn from the completed project. Whereas the initial workshop and follow-up workshops had to do with engendering team spirit and improving the performance of the project *in-hand*, the final workshop is aimed at consolidating the ground for continued partnering relationships between the various team members. It is also the opportunity to return to the SWOT analysis of the initial workshop. The contributors to the final workshop should analyse objectively what went well and what was less successful in terms of both the project outcome and the partnering process. If the formal evaluation procedures (see implementation stage) have been assiduously carried out for the duration of the project, then the progression of the project will have been fully charted (warts and all), and this will provide a good objective focus for the final workshop.

The participants at the final workshop should ideally comprise the participants from the first workshop plus the key stakeholders who joined the partnership team during the progression of the project, provided that the number of participants for the final workshop does not become too unwieldy. It has to be borne in mind that the final workshop is an important occasion for concentrated work, and may prove to be the stepping stone from single project partnering to strategic partnering. To this end, it is clearly of particular importance that the client body be as fully represented at the final workshop as at the initial workshop.

Strategic or multi-project partnering

There seems to be quite clear evidence that strategic, or multi-project partnering, will produce more substantial benefits than single project partnering[14,17]. The difference between single project partnering and strategic partnering is that strategic partnering has the added dimension of 'the development of a broader framework focusing on long term issues'[14].

Strategic or multi-project partnering occurs where the partnering team is in a position to enter into an undertaking for a series of projects, either new-build or a rolling maintenance agreement, and, as with single project partnering, can operate within conventional procurement agreements. Although there are significant benefits to be gained from strategic partnering, there are also significant hurdles to be overcome in its implementation, particularly with respect to national and international laws relating to free trade. For example in the United States, partly as a result of the anti-trust laws, over 90% of all partnering projects are of a single project nature[17] and, whilst the Central Unit on Procurement of the UK Treasury predicts a significant increase in partnering in the next few years, there is an implied caveat in its statement that 'Government policy is to use competition to achieve best value for money and to improve the competitiveness of its suppliers. *But* (our italics) the Government also appreciates the benefits of partnering and, in appropriate circumstances, these arrangements are being actively promoted'[14].

Despite these free trade reservations on partnering in general and strategic partnering in particular, it could be said that strategic partnering has all the advantages of single project partnering, only more so. The longer term nature of strategic partnering allows, amongst other things, for the development of a mutually beneficial physical infrastructure, for example, shared office accommodation and electronic data interchange (EDI). Although it is quite possible to create these arrangements for a single project, in highly technical, and high cost developments such as EDI, there are clear financial and technical benefits to be had in the semi-permanent relationship of strategic partnering.

It would appear that the attendant risks with strategic partnering are no higher than those associated with single project partnering, although in many cases, but not always, the magnitude of the turn-over and workload will be significantly higher in strategic partnering. It is worth noting that for any partnership to work, the partners must bring complimentary skills to the partnership. There is no point in the partners bringing the same or similar strengths, nor should one partner be motivated by the notion of eventual take over of the other partner. These points are valid for both forms of partnership but are particularly pertinent to a strategic partnering relationship.

Most strategic partnerships will come about through experience gained on single project partnerships. It follows therefore that the stakeholders in a strategic partnering arrangement are likely to be experienced in the mechanics of partnering. It is highly probable that the strategic partners will already have been involved with one another on individual projects. Strategic partnering operates at the macro level and single project partnering at the micro level. Kubal[10] uses the term 'second-level partnering' to describe long term or strategic partnering, demonstrating second level partnering by drawing on examples from the manufacturing sector. Here, established partnering arrangements between manufacturers and suppliers employ the 'just-in-time' inventory systems resulting in the supplier ensuring a steady workload, the manufacturer reducing production costs and the customer receiving goods at lower cost and higher quality.

Whilst the goals and objectives of strategic partnering are empathetic with the goals of project partnering they are not the same. Although terms such as time, cost, quality service and value can equally be applied to project and strategic partnering, the context and timescale in which they are applied is different. Similarly although, for example, the mechanisms for strategic partnering appear to be similar to those of single project partnering, with the use of initial and follow-up workshops using an external facilitator, the timetable for these workshops may be quite different, with a period of say 6 months elapsing from partner selection and the start of the first project of the strategic partnering relationship. Moreover 6 to 18 months may elapse from the time that management decides to explore strategic partnering until the commencement of the first project[14].

The aims and objectives of strategic partnering are encapsulated in the signing of a partnering charter, on similar lines to the single project partnering charter. However whereas the signatories to the project charter may number some 25 or so participants, the signatories to a strategic partnering charter will be restricted to one per partner. The charter will reflect the values, beliefs, philosophy and culture of the partners and these attributes will normally be translated into a charter in the form of a mission statement followed by a series of objectives.

A typical strategic partnering charter would be as shown in Figure 4.4:

Strategic Partnering Charter

Mission statement

Our mission is to work towards continually improving the spirit of trust and business relationships between our organisations so that we produce a better quality environment for our end users.

Objectives

To ensure value for money for our buildings users

To maintain a spirit of trust in all our relationships and co-operative activities

To demonstrate a genuine concern and interest in the productivity of all stakeholders

To produce buildings of excellence

To produce our buildings within time, on budget with zero defects

To ensure a reasonable profit for all the stakeholders

Company A...........................

Company B...........................

Company C...........................

Company D...........................

Figure 4.4 Strategic partnering charter

As can be seen from the strategic partnering charter, its aims and objectives are generic rather than project specific. Some charters may include a statement to the effect that 'nothing in this statement constitutes a partnership or binding agreement'[14] to emphasise the non-legal nature of the agreement. Given that this charter will be displayed in a variety of office and site locations, it may be felt that stressing the non-legally binding nature of the charter is a redundant statement which detracts from the spirit of trust expressed in the strategic partnering charter.

Legal and contractual implications of partnering

[Note: The legal and contractual implications of partnering will vary from country to country depending on the legal system in place. The following comments are made mainly in the context of the UK, Australia and New Zealand where the legal systems have a common basis, although even between these three countries there will be significant differences in terms of the trade practices in operation in each country. The added dimension of European Competition Law also affects the UK.]

Making explicit statements that a partnering charter does not create a legally binding relationship between partners, does not necessarily mean that none exists. Certainly the strength of the relationship is much less than in a true partnership. In this latter case each partner is considered to be the agent of the partnership and therefore, can bind the partnership with respect to third parties (a fiduciary duty)[14]. However as the Construction Industry Institute Australia (CIIA) report on partnering observes:

'Although the construction contract provides a framework of rights and obligations, partnering has the potential to impact upon the allocation of risk established by that contract and subsidiary contracts. If the partnering arrangement breaks down, a party may find itself in a position where it is necessary, or at least attractive, to assert that the contractual risk allocation has been altered, either by the provisions of the partnering charter or by subsequent conduct or representations in the course of the partnering process. This is, potentially, the major risk to partnering in Australia[24]'.

It is not our intention in this text to explore the legal ramifications of partnering, particularly since these will vary from country to country. However it is worth stressing that partnering relationships whether strategic or single project will have, if not ramifications, at least implications in any country in which it is practised. The following example, extracted from the CIIA report[24] illustrates this by listing the ways in which the partnering process could impact on the underlying construction contract :

- The implication of a contractual duty of good faith
- The creation of fiduciary obligations
- Misleading and deceptive conduct
- Promissory estoppel and waiver

- Confidentiality and 'without prejudice' discussions.

With respect to 'good faith', the CIIA report makes the following observations: 'Although Australian courts have, as yet, not been prepared to uphold a duty to perform a contract in good faith there is an indication that given a breakdown in the partnering arrangement, ...it would be open to a party to argue, that as the fundamental characteristics of partnering are consonant with good faith, and it is a US concept supported by the duty of good faith applied by the courts in that jurisdiction, there should be implied into its contractual relationships a general duty to act in good faith. Thus partnering could give rise to a situation where the court is prepared to recognise the concept of good faith; and therefore a party's rights to exercise its strict contractual powers might be limited by the duty to act in good faith". It should be noted that good faith is not the same as the duty to act reasonably. It is a broader subjective concept. Reasonableness, in legal terms, is a narrower objective standard. The report goes on to say that 'Finally, it may be open to a party to argue that the partnering charter forms part of the contractual documentation, even if there is no reference in the construction contract to partnering, and therefore, that the charter provided for the express duty to act in good faith in the performance of the contract."

This legal opinion is given by way of illustrating the potential contractual and legal risks of partnering. (For a more detailed discussion of this aspect of partnering readers are referred to the CIIA report *'Partnering: Models for success. Research Report 8'*, with the caveat that this report refers specifically to partnering in an Australian context.) We are not inferring that partnering is a legal minefield. However it is sensible to attempt to minimise any attendant risks, both legal and contractual, in a partnering agreement by taking precautions and anticipating potential legal and contractual issues during the initiation stage of the partnering process. CIIA recommend that the parties to the agreement should incorporate provisions and procedures into the partnering charter as follows:

- **Good faith**- by express provision clarify the issue for the partnering arrangement so that it will not be implied into the contractual arrangement.

- **Fiduciary obligations**- (an obligation to act to a higher standard of conduct than a simple commercial relationship, a feature of a true partnership) By express provision exclude fiduciary obligations arising, thus permitting the parties to freely pursue their own interests; or alternatively, limit the scope of the obligations to the purposes of the partnering process, thus permitting the parties to pursue their own interests outside the scope of their obligations.

- **Trades Practices Act**- it is very difficult to disclaim liability for misleading conduct by reference to a documented agreement.

- **Promissory estoppel and waiver**- (concerned with enforcing representations or promises as to future conduct, including promises not to rely on a party's strict legal rights, in circumstances where it would be unconscionable to do so). Incorporate into the partnering charter a procedure which must be followed if a party is to be denied its right to insist on enforcement in accordance with the construction contract's terms and conditions.

- **Confidentiality and 'without prejudice discussions- in relation to confidentiality** - include a confidentiality clause protecting confidential information by prohibiting it from being disclosed to parties other than the participants and for purposes other than the partnering process. In relation to the 'without' prejudice discussion, provide that any disclosure or concession made during the partnering process is made for that purpose, including any formal proceedings.

The CIIA report recommends that these provisions be incorporated into the partnering charter. It would appear to be difficult to incorporate these provisions into the partnering charter itself given the need for the charter to be an unambiguous statement of trust and co-operation which will be displayed in a variety of locations. It would be an anathema to the partnering process to have a set a legal qualifiers in small print in the partnering charter. However the CIIA report gives sound advice which should be heeded. The solution is perhaps to document the types of issues listed in the CIIA report, not on the partnering charter itself, but, as supporting documentation. The advice given in the CIIA report is not a return to the adversarial climate of the traditional approach, but a sensible precaution to minimise the legal and contractual risks to the participants.

Those undertaking partnering arrangements in the UK must also be aware of European law as it affects trading within the European Union. *'Trusting the team: best practice guide to partnering in construction'*[14] gives a detailed account of contractual and legal issues as they affect the UK. Those readers with particular interests in UK partnering are referred to this report. The following statement extracted from the *Trusting the Team* report clearly illustrates the importance of being attuned to contextual contractual and legal issues :

"Partnering will always be unlawful in EU law if its effect is to discriminate against undertakings on national grounds, or breach the fundamental freedoms of the European Union - namely the freedom of movement of goods, services, workers and capital. Even if a partnering arrangement does not affect trade between member states, it may still be void, punishable by fine or liable in damages under English law if the agreement imposes restrictions on the price and supply of goods."

Dispute resolution

It would be something of a utopian ideal to assume that disputes will not arise, even in the non-adversarial culture of a partnering relationship, and, as with contractual and legal issues, it is prudent to have contingencies in place should the eventuality arise. In a recent survey of partnered projects in Australia[24], 91 per cent of respondents agreed that a dispute resolution plan was an essential element of a partnering arrangement. In projects where partnering arrangements had been unsuccessful, 43 per cent of these had no formal dispute resolution plan in place.

Although it is important to have dispute resolution procedures in place, it must be acknowledged that the incident of disputes is likely to be less in a partnering relationship, because the partnering implementation plan should incorporate mechanisms preventing escalation of differences into full blown disputes[25].

The following issue resolution process is a typical approach[26]:

- Resolve the problem at the lowest level of authority
- Unresolved problems should be escalated upward by both parties in a timely manner, prior to causing project delays
- No jumping of levels of authority is allowed
- Ignoring the problem or 'no decision' is not acceptable

'Partnering calls for providing as many opportunities for communication as possible. Traditionally a velvet curtain is often used by project team members to hide problems and hope they can resolve it themselves...Through an issue resolution process the stakeholders' experiences both good and not so good, are put on the table. Risks and potentially difficult areas of the contract can be discussed openly. A high-trust culture is developed where everyone feels free to express ideas and make contributions to the solution.'

Conclusion

There may be many in the construction industries in the UK, Australia and New Zealand who view any management concept which emanates from the United States with a certain degree of scepticism, particularly if, as is the case with partnering, the concept is being promoted with a certain amount of evangelical hype. The promise of a win-win business environment may seem, to some, to be an unachievable utopia. However there are well documented case studies[14,24,27] of many successes in the use of both project and strategic partnering. Bennett and Jayes[14] make the following comments: " The initial investment in workshops and careful selection of partnering firms rapidly translates into significant net benefits that on individual projects can amount to 10% of total costs. Over time, strategic partnering can achieve the whole of Latham's 30% cost saving target."

The CIIA[24] survey of partnering in Australia found that "nearly 85% of respondents would undertake another partnering project".

'Not surprisingly, partnering was considered a great success by the respondents who had experienced a partnered project with no contractual claims. Conversely, partnering was considered a failure when contractual claims amounted to more than 5%. In 56% of cases where partnering was a success there had been repeat business between partners. Where partnering had been unsuccessful there had been no repeat business." It is worth tempering enthusiasm for partnering with the findings of the CIIA survey which indicated that although partnering was seen as successful by the respondents and that those who had experienced successful partnering would partner again, on a five point scale measurement of partnering success the mean of 3.83 was "not overwhelmingly high".

Given that partnering has to do with human relationships and trust, it can never be infallible. However the balance of evidence points in favour of partnering as a construction management concept which is capable of achieving a cultural shift from adversarial to non-adversarial, and hence it has the potential to significantly improve the net benefits for all the stakeholders. The principal benefits of partnering are generally seen as being[24]:

- Reduced exposure to litigation
- Improved project outcomes in terms of cost, time and quality
- Lower administrative and legal costs
- Increased opportunity for innovation and value engineering
- Increased chances of financial success
- A more detailed set of benefits was identified in the CIIA[24] study and was ranked in order of importance as follows:
- There was exchange of specialist knowledge
- Reduced exposure to litigation
- Lower administration costs
- Financial success because of win-win attitudes
- Positive effect on claims costs
- Positive effect on schedule duration
- Better time control

- Better quality product
- Prompted technology transfer
- Better cost contro
- Fostered innovation
- Reduced rework
- There was evidence of innovation and improvement
- Improved safety performance
- More profitable job
- Fewer errors in documentation
- Innovation as a result of information exchange
- Mechanism for recording innovation

In the introduction to this chapter we began with the comment that "Partnering is more than simply formalising old fashioned values, or a nostalgic return to the good old days when a 'gentleman's word was his bond'. For partnering to be successful the ideals of partnering have to be implemented through the planning and implementation mechanisms which we have described. "Critical performance indicators have a key role in this respect. Constant monitoring of performance is one of the most critical factors in achieving partnering success"[24].

In addition to the need to maintain good lines of communication, it is important that communications are diligently recorded so that all parties to the partnership are fully informed of the ongoing situation. Partnering is however not just about high ideals and creating an atmosphere of mutual trust, it also involves having good organisational systems in place and emphasising tasks of a routine nature, such as the keeping of proper records. For example respondents to the CIIA study rated the keeping of minutes at partnering meetings as the most important feature of continuous evaluation.

The concept of partnering is analogous to the torch of the Olympic movement. As the torch is passed from hand to hand, there is the ever present danger that the torch will be dropped and the flame extinguished. Similarly, if the ideals of partnering are not continually revisited and carefully nurtured, then the partnering message is likely to become distorted in the process of dissemination.

---

[1] Hellard R.B. (1995) *Project partnering: principle and practice*. Thomas Telford, London.

[2] Schultzel H.J. and Unruh VP. (1996) *Successful partnering: Fundamentals for project owners and contractors.* John Wiley, New York.

[3] Hinks A.J., Allen S., and Cooper R.D. (1996) Adversaries or partners? Developing best practice for construction industry relationships. In: *Langford DA, Retik A, editors. The organisation and management of construction; shaping theory and practice. Proceedings of the CIB W65 International Symposium*; Aug. 28 - Sept. 3; Vol. 1, 220-228, Glasgow, Scotland.

[4] Dubbs D. (1993) Partnering means making friends not foes. *Facilities design and management*, June; p48.

[5] Wright G. (1993) Partnering pays off. *Building design and construction*, April; p36.

[6] Kliment S.A. (1991) Partnering makes sense. Architectural record, March; p9.

[7] Stasiowski F.A. (1993) Partnering - the only approach for the 90's, *A/E Marketing Journal*, Dec, p1.

[8] Hancher D.E. (1989) Partnering: *Meeting the challenges of the future.* Interim report of the task force on partnering. Construction Industry Institute, University of Texas, Austin.

[9] Godfrey K.A. Jr., (ed.) (1996) *Partnering in design and construction.* McGraw-Hill, New York.

[10] Kubal M.T. (1994) *Engineered quality in construction: partnering and TQM.* McGraw-Hill, New York.

[11] National Economic Development Office. (1991) *Partnering: contracting with out conflict.* June, NEDC, London.

[12] Cowan C., Gray C. and Larson E. (1992) Project partnering. *Project Management Journal*, Dec; 5-21.

[13] Gyles R.V. (1992) *Royal commission into productivity in the building industry in New South Wales*: Government of New South Wales, Sydney.

[14] Bennett J. and Jayes S. (1995) *Trusting the team: the best practice guide to partnering in construction.* Reading Construction Forum, Reading.

[15] New South Wales, Department of Public Works and Services. (1995) *Contractor accreditation scheme to encourage reform and best practice in the construction industry:* Government of New South Wales, Sydney.

[16] Anonymous. (1995) Partnering and mediation gaining widespread acceptance, says survey. *Civil Engineering* May; **65**(5); 28-30.

[17] McManamy R. (1994) CII Benchmarks savings. ENR Aug. 15.

[18] New South Wales, Department of Public Works and Services. (1995) *Capital project procurement manual:* Government of New South Wales, Sydney.

[19] Stevens D. (1993) Partnering and value management. *The Building Economist*. Sept. 5-7.

[20] Allot K. (1991) Partnering - an end to conflict? *Process Engineering* Dec; **72**; 27-28.

[21] Katzenbach J.R. and Smith D.K. (1993) *The wisdom of teams: creating the high-performance organisation.* Harvard Business School Press, Boston, Mass.

[22] Moore C., Mosley D. and Slagle M. (1992) Partnering: guidelines for win-win project management. *Project Management Journal* Mar; **23** (1), 18-21.

[23] Kanter R.M. (1989) Becoming pals: pooling, allying and linking across companies. *Academy of Management Executive* **3**, 183-193. In: Cowan C, Gray C and Larson E. (1992) Project partnering. *Project Management Journal*, Dec; 5-21.

[24] Construction Industry Institute Australia (CIIA) (1996) *Partnering: Models for success.* Research Report 8. Construction Industry Institute Australia.

[25] Stephenson R.J. (1996) *Project partnering for the design and construction industry.* John Wiley, New York.

[26] Construction Industry Development Agency (CIDA). (1993) *Partnering: A strategy for excellence.* CIDA/Master Builders Australia.

[27] Warne T.R. (1994) *Partnering for success.* ASCE Press, New York.

5 Value management

Introduction

Most people would agree that there is a difference between cost and value and that of the two value is much more difficult to define. The value of a child's battery operated radio in a modern home stashed with satellite TV and sophisticated hi-fi may be very small. When a snowstorm however cuts the electricity supply and the small battery operated radio is the only means of hearing news, its value is drastically increased.

Even Chambers English dictionary[1] has several definitions of value ranging from *that which renders anything useful or estimable* to, more simply, *price*. It is because of these different concepts embodied within the word value that value management is sometimes difficult to understand and it is possibly for this reason that it has been confused with cost saving, buildability and cost planning[2]. Such confusion has, not surprisingly, led to a fragmentation in the use of value management and in today's construction industry the technique is at the crossroads of its development[3]. Following an initial flurry of enthusiasm in the late 1980s, use and interest in the technique diminished, largely because there was a lack of appreciation of the different concepts that can exist within in the term value. This can be clearly seen in the historical development of value management from its invention in the 1940s to its use in today's construction industry.

Historical development

Value analysis was the term first used by Miles[4] to describe a technique that he developed at the General Electric Company during the Second World War. The technique began as a search for alternative product components, a shortage of which had developed as a result of the Second World War. Due to the War however, these alternative components were often equally unavailable. This led to a search not for alternative components but to a means of fulfilling the **function** of the component by an alternative method. It was later discovered that this process of function analysis produced cheaper overall products without reducing quality and after the War the system was maintained as a means of both removing unnecessary cost from products and improving design.

The central feature of Miles' work was definition of all functions that the customer requires of the product. These functions were defined only in terms of one verb and one noun. Miles believed that if such a definition could not be achieved then the real function of the item was not understood. The defined functions were then evaluated in terms of the lowest possible cost to achieve them and this evaluation was then used as a means of finding alternatives that also fulfilled the functions. Miles illustrated his work with an example of an electric motor screen which needed to perform the functions is shown in Table 5.1.

Table 5.1 Function definition

	Verb	Noun
1	Exclude	substance
2	Allow	ventilation
3	Facilitate	maintenance
4	Please	customer

These functions were then evaluated by selecting the cheapest possible means of achieving them. These were as follows:

- The function of 'exclude substance' was evaluated on the basis of the cost of a sheet of metal to shield the motor.

- The function of 'allow ventilation' was based on the additional cost of putting holes in the sheet metal.

- 'Facilitate maintenance' was evaluated by adding the cost of a spring clip to allow the sheet metal to be removed.

- 'Please customer' was based on the cost of painting the metal.

- The costs of these are shown in Table 5.2.

Table 5.2 Allocating cost to function

	Verb	Noun	Cheapest means of achieving function	Lowest cost to achieve function
1.	Exclude	substance.	Sheet metal	$0.15
2.	Allow	ventilation.	Holes in metal	$0.15
3.	Facilitate	maintenance.	Spring clip	$0.10
4.	Please	customer.	Paint metal	$0.10
Total lowest cost to achieve all functions				$0.50

This total lowest cost can therefore be viewed as the true value of the screen, since it is only this cost that needs to be incurred in order to meet all the functions that the customer requires. In going through this process Miles costed the functions of the screen based on the lowest possible cost of achieving them; in this case $0.50. This lowest cost was then compared to the actual cost of the existing screen: $4.75. This clearly highlighted that much of the cost incurred in producing the screen actually achieved no function.

A by-product of defining and evaluating function is the ease with which it allows further means of achieving the function to be generated. This can be illustrated with a construction example. In a hotel development the architect had included a child's paddling pool next to the main hotel pool. The function of this pool was not, as might first be thought, to provide a leisure activity for children. It was in fact a safety measure to keep them out of the main pool. Once this function of the child's pool was defined as 'keep (children) safe' it was much easier to generate alternatives that satisfied this function. A play area, a small playground or even a crèche would all meet the requirements. In the end the design team settled for a water spray which fulfilled the function of safety for much less cost than the original paddling pool.

This function definition, function evaluation and generation of alternatives are collectively called function analysis and it is this basic technique which forms the basis of value engineering.

Clearly in complex organisations a system would be required to carry out function analysis: it could not be done on an ad hoc basis. When the function analysis would be carried out, who would do it, and how it would be organised would need to be considered. For this reason systems of value engineering as it now came to be called grew up around the technique of function analysis[5]. Figure 5.1[6] shows a typical system that developed. Central to the system was the concept of function analysis. The studies were organised into a series of steps known as the job plan and were carried out by a value engineering team at some time during the products life cycle.

Up until the 1970's value engineering was only being used in the manufacturing sector but at around this time it started to be used in the United States for construction projects. Use was mainly by government agencies[3] who for various reasons adapted the manufacturing value engineering systems and tailored them for their own use.

As shown in Figure 5.2 the 40-hour workshop became a feature of value engineering when it began to be used in the construction industry. In addition value engineering was carried out at the 35% design stage using a team external to the project; that is a team not forming part of the design team. It was largely this system of value engineering, imported from the United States, that was first used in the UK construction industry.

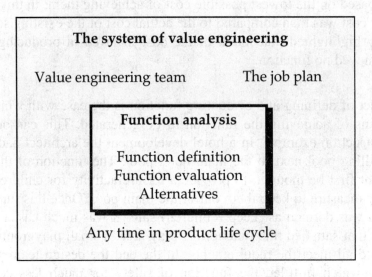

Figure 5.1 The system of value engineering in relation to function analysis

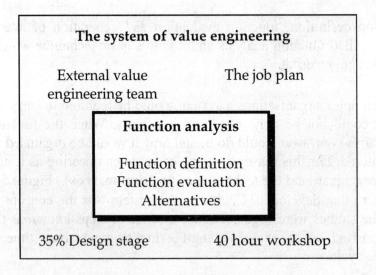

Figure 5.2 Systems of value engineering in construction

The problem with this American system, which is still widely used in the US, is that it deviates from the original work of Miles. It does not, other than in name, include function analysis. Value engineering in the US is basically a design audit. It consists of a 40-hour workshop structured loosely around a job plan. It is carried out at 35% design by an external team. It involves the selection of high cost areas and the generation of alternatives to them. The selection of high cost areas is a fairly loose procedure. It is based on the comparison of elemental costs with the cost of cheaper alternatives, along with a more general analysis of cost centres of the project. This nebulous approach results in a fairly broad value engineering output encompassing design changes and cost cuts from all disciplines. This output however cannot be attributed to function analysis. The actual workshop itself as an autonomous unit is a critical contributory factor in the success of value engineering studies. Within the workshop, the degree of success of the study relates largely to the personalities involved, particularly that of the leader, the timing of the study, the interaction of the value engineering team, the input of the design team and the role of the client. The technique of function analysis bears little or no relationship to the output of the study.

When British companies tried to use the US system of value engineering they found that they couldn't make it work. The reasons for this were numerous but largely related to the original objectives of the value engineering studies. The US system of value engineering was born out of a need for greater accountability on government projects. (Almost all value engineering activity in the US is government work.) The situation in the UK was very different. The quantity surveying system provided all the accountability that was needed. Value engineering was required to provide a platform for the examination of value as opposed to cost. In this light it was hardly surprising that the US system of value engineering broke down in the UK.

Once this breakdown had occurred the UK construction industry was faced with two options. It could abandon value engineering altogether or it could go back to the original work of Miles and build its own system that satisfied its own objectives. It appears that the latter route has been chosen and that new systems are developing under the title of value management. For this reason only this terminology will be used in the text from this point onwards. What this choice means is that the UK is essentially starting with a clean sheet in building new systems of value management. However as mentioned above the US experience has shown that in addition to function analysis there are other components of value management systems, such as the team or the timing of the study, that influence the success or otherwise of value management. To build an effective value management system these components need to be investigated and the next section of this chapter examines these components in detail.

Before we move on a final point regarding the American system needs to be clarified. The authors have defined value management as incorporating function analysis. This is our definition. As Chambers dictionary illustrated the concept of value may cover many things. The US system, despite neglecting function analysis, is useful and fulfils the objectives for which it is required.

Function analysis

On a recent visit to my new local surgery the nurse complained about the construction of the facility. In the room in which we were sitting the door was directly opposite the work cabinets and this meant that the patients trolley bed could not be pushed into or out of the room. It was necessary for a patient to be wheeled in on a wheelchair and then lifted onto the trolley. It would, she complained, have been much easier if the door had been placed on the other side, allowing the trolley to be moved in and out freely without obstruction from the work cabinets.

Clearly in designing the room the designer did not consider what the room would be used for. He did not know that the room would be used to examine patients who were incapable of walking into the room and climbing onto the bed themselves. That is, he did not establish the function of the room correctly. The most likely reason that he did not do this was because he failed to involve in the design process the people who would use the room. As a result the value of the room was diminished, in that it did not properly perform the function for which it was required.

This failure to provide buildings or parts of buildings which properly perform their functions is a common problem in the construction industry and it is this that forms the basis of function analysis. In the context of function analysis it is assumed there is a close relationship between the provision of function and the achievement of value. Where all functions are accomplished at the lowest achievable cost there is good value. Where no function is achieved, or where function is achieved at too great a cost, there is little or no value.

The problem that arises at this point is that even if the designer can define the functions that a particular item is required to perform, how can he know that the functions are in fact being provided at the lowest achievable cost? As explained function evaluation which takes place after function definition, assesses the defined functions based on the lowest possible cost to achieve them. As an example think about a window, which in a particular location may have the sole function of allowing light to enter (admit light). The cheapest possible means by which this can be achieved is by forming an opening in the wall. The cost of doing this is therefore the **value** of the function. Naturally this example must not be taken literally. It is highly unlikely that the sole function of a window is to admit light. The example however illustrates that anything spent above the cost of achieving the function is unnecessary and reduces the value of the item. Practical illustrations of this example can be seen in many hot climates where the function of admitting light is in fact achieved solely by forming a hole in the wall.

A further benefit of function evaluation is that it is creative. It acts as an aid to finding other solutions that also satisfy function. Looking for the lowest possible cost to achieve function tends to lead to the question 'if a hole in the wall will satisfy the function what else will?' A hole in the roof, a see-through wall, or an electric lamp will all equally satisfy the function of admitting light.

As summarised in Figure 5.3 the three items of function definition, function evaluation and creative alternatives are collectively called function analysis and it is this that forms the basic core of value management. Without this technique then any exercise that is carried out on a project, useful though it may be, cannot be described at value management. For this reason each stage of the technique is examined in greater detail.

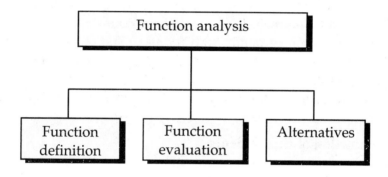

Figure 5.3 The stages of function analysis

Function definition

Function is defined through the use of a verb and a noun because this clarifies the function. Generally if the function of an item cannot be defined in terms of a verb and a noun then its function is not fully understood.

The problem that arises with function definition in construction is one of level. In the example the function of an element of the building, the window, was defined but it is also possible to break the window into a series of components which could also be defined, e.g. the head, sill or locks. Equally the function of the spaces provided by the project could be defined. In a school, for example, this might include the functions of classrooms and staffrooms. Alternatively it is possible to step back still further and define the function of the school as a whole. Defining the functions at these four different levels would produce four different results. Which one of these levels should therefore be chosen?

Defining the function of the project as a whole

Many projects are not what they seem. As an example think about the functions of a bridge, the most obvious function of which is to 'convey traffic.' However this may not necessarily be the case and the real function may be to take industry to another part of the city, or to provide work for the construction industry or to reduce traffic congestion. Correct definition of the project function at this level can be useful because if the correct function is, for example, to move industry to another part of a city, then it is clear that this can be achieved other than through the construction of bridge. The offering of grants or subsidies is an obvious alternative.

Defining the functions of the spaces within the project

Once it has been decided that a particular project is required, the functions of the spaces within it can be defined. In a school for example we can ask what is the function of classrooms, playgrounds, staff rooms or sports facilities. The function of the playground might be to give the children an opportunity to let off steam, to allow them some fresh air, or to give the teachers a well earned break. Depending on the function, the design solution, and any viable alternative to it, will vary.

Defining the function of the elements

In addition to project and space function, the function of project elements can also be defined. Think once again about a window. If you were asked you would probably define its functions as 'provide ventilation' or 'admit light'. But now think about the function of a window in a prison cell. In this case the function is not to admit ventilation as the window will rarely, if ever, open. Nor will the function be to admit light, as often the amount of light provided is insufficient even for day use. The real function of the window is to 'humanise environment'.

The definition of elemental functions can, therefore, also provide a clearer insight into design and design alternatives. However it is not possible to define the functions of elements generically, without reference to the building itself. It can never be assumed that the function of internal walls is to 'divide space.' The function of internal walls in a prison cell is entirely different from the function of internal walls in a toilet block. Neither have the function of dividing space.

Defining the function of components

It is possible to divide elements onto components and once again this can provide insight into design and design alternatives. In breaking a window into components of sills, heads, and jambs, these parts can be examined to see how well they meet the required functions.

Which level of function should be selected?

Given that there are four possible levels of function definition which one should be used for a construction project? The answer to this depends on the client, the level of design completed and the project generally. Where the client is adamant that the project concept is the one required then there is little purpose defining its function and a function definition on space may be the most appropriate. Likewise if the spaces provided are fixed then an elemental function definition may be the best way of improving value. Generally in construction function definition is not carried out on components as this is viewed as the remit of the manufacturer.

It is of course possible to do all three or any two of these function definitions at the same value management study. However this may be too complex and time consuming, particularly when faced with very large projects. Figure 5.4 summarises the position with regard to function definition. The general rule is that the higher the level of function definition the greater the capacity for changing the project and therefore the greater the potential for improving value. The highest level of function analysis possible should always be selected.

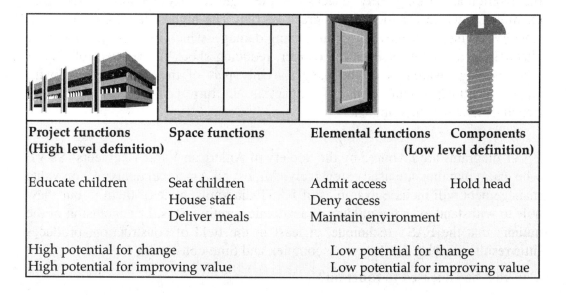

Project functions (High level definition)	Space functions	Elemental functions	Components (Low level definition)
Educate children	Seat children House staff Deliver meals	Admit access Deny access Maintain environment	Hold head
High potential for change High potential for improving value		Low potential for change Low potential for improving value	

Figure 5.4 Levels of function definition

Function before or after design

High level function definition is much more beneficial to the project. It therefore follows that the ideal time to do function definition is before any design has taken place, thereby maximising the potential for change and for improving value. However it is more often the case that some design exists. The problem with this is that the value management team will tend to use this design to define functions.

What this in effect means is that they are accepting the functions as designed before they have established that those functions are required. After all function is based on the requirement of the customer and user, so strictly the functions that come from them will not change, regardless of the presence of an existing design. This may seem like a subtle distinction but it is an important one. Too many design decisions are based on an acceptance of what was done in the past. Often however it is better to start with a fresh approach based on clients needs only. In that respect only necessary or desired functions will be included.

FAST diagrams

The problem of different levels of function definition was also recognised by the American value managers[7] who tried to show the inter-relationship between these levels through use of a function analysis system technique or FAST diagram[8] such as Figure 5.5. The diagram operates by starting with the primary function or high level function and asking of it 'how is this to be achieved?' The answers to this question form the next layer of functions and the question of 'how' is also asked of them until the final function is reached. Operating the opposite way on the diagram and asking 'why' checks that the logic of the diagram is correct. The example shown in Figure 5.5 is a crash barrier. The highest level function is to save lives which is achieved by minimising damage, which in turn is achieved by channelling traffic away from the danger, reducing shock in the event of impact and ensuring awareness of the danger. (Only part of the diagram is shown.) Reducing shock in the event of impact is, in turn achieved by absorbing, transferring and redirecting energy.

FAST diagrams are favoured by the Society of American Value Engineers (SAVE) who claim that they are used extensively. Almost all American textbooks on value management will include examples of FAST diagrams, none of them in our view able to withstand even the mildest of academic scrutiny. It is the conclusion of the authors that the FAST technique, at least in the field of construction, produces little results for what can be a very complex and time-consuming exercise.

Can cost be allocated to function?

Some value management practitioners, on completion of function definition, allocate the estimated cost of the project among the functions[7]. If for example the main functions of a hospital were defined as follows:

- Treat patients
- Diagnose patients
- Allow stay

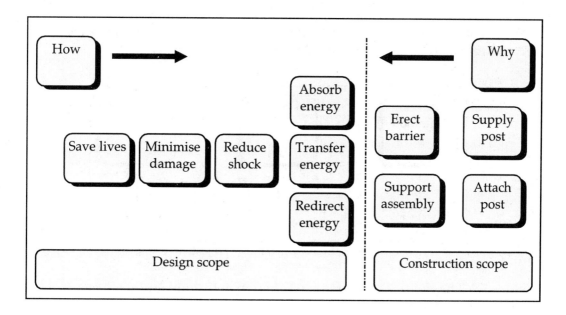

Figure 5.5 A FAST Diagram[9]

Then the estimated cost of the project is divided up between these functions so that the client can see what it is costing him to provide each of the main functions. This is the method used by Howard Ellegant[10] along with other prominent US value management practitioners. It has some merit in that a client may not want to include a particular function once he becomes aware of the financial implications of providing it. On the negative side however it is very difficult to allocate cost to function, since some items such as the foundations or central plant must be included regardless of whether a function is included or not. The other problem with allocating cost to function is that it really compares apples with pears. We have already seen that function definition should not relate to the existing design, it is best carried out independently. If this is the case then there is no relationship between the function definition and the estimated cost, since the latter is only a reflection of the existing design (Figure 5.6). For this reason it is the authors view that allocating cost to function is largely pointless. However, it is appreciated that there is an argument in favour of it.

Function evaluation

Once functions are defined they are evaluated based on the lowest cost to achieve them. In the example used earlier we considered a window with the sole function of admitting light. The cheapest way of achieving this function is to put a hole in the wall. This solution is not to be taken literally; it is merely a way of illustrating the value of the function and also a means of generating alternatives that also meet the function. In reality we do not need to accurately price this hole in the wall but just have some idea of its cost. The purpose of function evaluation is a catalyst to creative alternatives. It does not need to be a literal costing exercise.

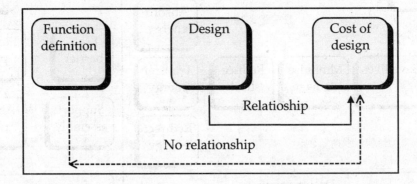

Figure 5.6 The relationship between function definition and project estimate

It is possible therefore that a value management study will produce two sets of costs; estimated cost of the project allocated between the functions (function costs) and also the lowest cost to achieve those functions. In some value management studies a worth factor is calculated by dividing function cost by the lowest cost to achieve function in order to arrive at a mathematical representation of worth (Figure 5.7). These factors can give some indication of the unnecessary cost being spent on functions and may be used to prioritise functions that need to be re-examined. Alternatively no cost is allocated to function and the function is viewed only in relation to the lowest possible cost to achieve it (Figure 5.8). These two approaches to function evaluation are summarised below. The authors recommend the latter approach.

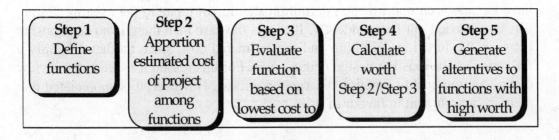

Figure 5.7 Using estimated costs to evaluate function

Alternatives

Once functions have been evaluated the evaluation can be used to generate alternatives by asking "*what else will also achieve the function?*" Going back to the example of the window in the prison cell the cheapest way to 'humanise environment' may be to provide a radio. But what else will also do the job? A TV, a mobile library, bigger cells. All of these may 'humanise environment' and they may do it in a better and cheaper way than a window.

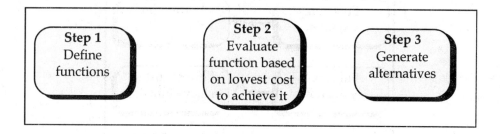

Figure 5.8 Evaluating functions without the use of estimated costs

The lowest cost to achieve function acts as a catalyst in generating alternatives that satisfy function it may still be difficult to get individuals present at the value management study to put forward creative ideas that also meet function. For this reason brainstorming is often used to assist with the generation of alternatives. Brainstorming operates by encouraging wild and outrageous suggestions in the hope that these will generate good and workable ideas. The advantage being that because ridiculous suggestions are encouraged and welcomed and there is no embarrassment attached to them or any good ideas that follow. The most stupid idea is viewed as the best one. Participants are encouraged to shout out the ideas as they come to them and formality is discouraged. (Some authors go so far as to suggest alcohol as an aid to this stage of the study[9].) Brainstorming is a general management technique, it is not peculiar to value management. Much has been written about it but we only have enough space in this text to outline its general rules[11]:

- No criticism of ideas is allowed until all ideas have been collected.
- A large quantity of ideas is required.
- All ideas are recorded.
- The best ideas often come from the inexperienced.
- It is a group exercise and ideas should be built on, and used to spark other ideas.

Figure 5.9 summarises function analysis; the core of value management. It is a three phase approach of function definition, function evaluation and generation of alternatives. Function analysis is achieved through the verb noun definition; function evaluation is through the evaluation of the lowest cost to achieve function and alternatives are created through brainstorming.

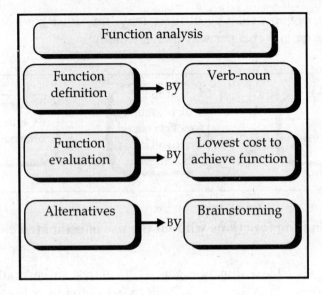

Figure 5.9 Summary of function analysis

Function analysis is not easy to implement and may take years of practice to refine, it cannot be learned only from the pages of a book. In addition it is not a stand alone technique. In order to develop a value management **system** we need not only function analysis but an effective means of carrying it out. How to organise the study, who should do it and when it should take place all need to be carefully considered. The next section of this chapter deals with this additional components of the value management system.

Organisation of the study

An integral part of value management is the job plan which is a five stage process of organising a study. The five stages are outlined below.

Information stage

At this stage all the information required for the project is gathered.

Analytical phase

At this stage the function analysis takes place in two sections.
- Define the function
- Evaluate the function

Creative phase

At this stage the function analysis is used as a basis of generating alternatives that meet the functions as defined in the analytical stage.

Judgement phase

At this stage all the possible alternatives that meet function are evaluated.

Developers' phase

The ideas that are considered worthy of further consideration are selected and developed further.

Who should carry out the study?

The purpose of value management is the optimisation of the needs and wants of those with an interest in the project. It therefore follows that in order to be successful, value management must be carried out as a team exercise. There is really no exception to this and right from the beginning of value management in the 1940's it has always been viewed as a team approach. There is much that has been written on group dynamics and leadership in the fields of value management[12] and general management and anyone who wishes to study it in depth will find a dearth of information. However for those interested in the practicalities in relation to value management the following is a summary of the major issues.

Who should constitute the team?

The value management team can constitute the design team or an external team that are new to the project and who have had no previous involvement in it. It may, alternatively, be a mixtures of the two, either with or without the presence of the client. In addition specialists may be invited if the project has particular problems and a specialist input is required.

The question of which is the better way of carrying out the study; the design team or an external team, is open to dispute. In the United States the preferred method tends to be an external team, whereas the design team is preferred in the UK. There is no wrong or right answer and it is ultimately the choice of the client. The following is a summary of the advantages and disadvantages of each method.

Advantages of using an external team

- Objectivity

- The team can be selected for their particular skills whereas the design team is already established

- Re-assurance for the client that the design produced by the design team is a good one.

The disadvantages of using an external team

- The design team may have difficulty accepting the presence of an external team.

- The design team has already formed and overcome many of the teething problems that groups experience. An external team will take time to come together.

- An external team may not really be objective at all. In some cases, particularly in the case of large clients, the external team may want to 'put on a good show'. One way of doing this is to be over critical of the project under study; implying that *they* would have done it better.

- Using an external team is more disruptive than using the design team.

- If function analysis is being used correctly then an external team is not really necessary.

- The external team may be overly concerned with showing a cost reduction and may reduce specification as a means of achieving it, without any real regard for value or function.

- The external team lacks depth in their understanding of the project.

- The external team is expensive.

- An external team raises problems of design liability if they make any changes to the design which later prove defective.

There is of course no such thing as the perfect team. Despite the longer list of disadvantages there are undoubtedly times when an external team may be advantageous - when a project has serious political difficulties for example. However it is the authors' view that in the majority of cases the design team will provide a more effective value management study than an external team.

Who should lead the team?

The value management team leader, or facilitator as they are generally called, needs a skill base that includes an in-depth knowledge of function analysis, group and team building, evaluation of project alternatives and a knowledge of construction. It is unlikely that any member of the design team would have these skills and it will probably be necessary to call on an external facilitator. Value management is a technique built on experience and it is generally recommended that an external facilitator be used.

Should the client attend the study?

A value management study cannot be carried out without the presence of the client, even at the stage of elemental function analysis. The user would also be able to provide a useful insight into the functions of the building and should always be invited. Unfortunately this is rarely the case.

Team organisation

The value management team should not be allowed to become too big. There has, to date, been no research into the optimum size of teams for a value management study but anecdotal evidence suggests eight as a maximum.

All members of the value management team should be at approximately the same level of seniority, otherwise junior members tend to feel intimidated. In addition team members must have enough authority to make major decisions about the project.

The format of the study

Due to the early and fairly strong influence of the United States on the development of value management in the UK, the technique is often associated with the 40-hour workshop. This system operates by getting the value management team together for a period of 40 hours, away from the normal working environment, so that the study can be carried out free from interruption, thereby providing an environment suitable for the creativity needed to generate alternatives. The idea of the workshop has survived in the UK but rarely is 40 hours deemed necessary and evidence[3] suggests that two days is an average length of study in the UK. The reason for the 40- hour workshop in the United States was largely the efforts of the Department of Defence to standardise the value management procedure and neither SAVE nor any other body suggest it as the optimal solution. American 40-hour workshops are highly cost oriented and the value management team must produce full documentation of all proposals, along with a costing of them. It was the authors experience that at least half of the time in the 40-hour workshop was used in costing and writing up proposals; making them suitable for presentation to the client. In the UK where there is much less emphasis on cost, this type of exercise is rarely deemed necessary. The study can therefore be carried out much more quickly. As an interesting aside to this is that the US Department of Defence value management programme is so cost oriented that at the end of each study the savings proposed are divided by the cost of the study to give a return on investment!

As value management is a team exercise it is almost impossible to get away from the concept of the workshop. Some value management practitioners do claim to have an integrated system, value management taking place simply as a part of the normal design procedures. There are no evidences that such an approach could not work but in the opinion of the authors the technique of function analysis is aimed at changing mindset; as such an integrated approach is unlikely to succeed.

Where should the study be carried out?

There is no strict rule about where a value management study should be carried out but most value management consultants or facilitators recommend that it be away from the normal working environment, be in a hotel or conference facility.

The timing of the study

A value management study can be carried out at any point in a projects life cycle. The timing should not, in fact, make any difference to the outcome, since the functions of the project do not change simply because the design is at an earlier or later stage of development. However, the amount of redesign that may be required increases as the project develops. As a result the cost to change will also increase, as will the reluctance of the design team to make the changes. As outlined earlier function analysis can be carried out at various different levels (project, space and elemental) and naturally if a function analysis is required on the elements only, (possibly because the space allocation is fixed by planning permission or other constraints) it cannot take place until the appropriate level of design is reached.

Some value management writers suggest that the levels of function analysis corresponds with stages of design development and that value management can be carried out more than once[13], to correspond with the design stages outlined below.

Inception

Value management can be used as a means of deciding if the project is really needed. The highest level of function analysis is used. For example a local authority decides to build a new power station. A function analysis shows it is required because the existing power station cannot satisfy demand. The function of the power station is therefore to 'satisfy demand.' This can be achieved by either increasing the supply of electricity or reducing the demand. One way to reduce demand is to encourage people to use energy saving light-bulbs. Rather than build a power station it might therefore be better to give out free energy savings light-bulbs.

Brief

Once it is definitely decided to go ahead with a project, value management can be used to formulate the brief. Space level function analysis is used. Naturally it would still be possible to decide the project was not needed at this stage, but this would result in abortive design. Using value management at this stage as a means of formulating the brief, is the stage seen as most beneficial by some value management writers and practitioners[13]. For example a ward in an old people home has functions defined as 'allow stay', 'facilitate nursing', and 'provide food.' The provision of food then generates various alternatives such as self catering, meals on wheels, restaurant, import from catering or cook on the premises. All of these are viable alternatives particularly given the current emphasis by large organisations on outsourcing.

Outline proposals (35% design)

At this stage the design is more developed and it is possible for the value management team to offer alternatives based on the elemental design and specification. Elemental function definition would be used. Once again value management proposals based on changing the nature of the project or its spatial layout could be put forward but this would require redesign work at this stage.

During the construction of the work

Value management proposals put forward by the contractor have been used for a considerable period, particularly in the USA where these are known as value engineering change proposals (VECPs). Often the contractor is given some financial incentive to make the proposal and this is usually in the form of a percentage of the saving achieved. There are many problems associated with VECPs, not least the question of design liability. In addition many contractors operating under traditional procurement methods particularly feel no incentive to offer change proposals, since the saving achieved maybe outweighed by reduction in the scope of work.

How should alternatives be evaluated

The methods for evaluating value management proposals are numerous but perhaps the one most commonly used is the weighted matrix an example of which is shown in Table 5.3.[11]

Table 5.3 The weighted matrix

Method	Initial cost	Maintenance	Aesthetics	Energy impact	Construction time	Total
Weighting	6	3	10	3	3	
Surface mounted incandescent	18/3	3/1	20/2	3/1	6/2	50
Mercury vapour	18/3	9/3	20/2	6/2	6/2	59
Fluorescent fixtures	24/4	12/4	10/1	9/3	9/3	64

The matrix shows a range of possible light fittings along with some criteria against which they will be evaluated. Each of the criteria is weighted to reflect their importance, so that in the case above aesthetics is viewed as the most important criteria, cost as the next most important and so on. Each of the fittings is then given a rating for each of the criteria based on the following scale.

Excellent	5
Very good	4
Good	3
Fair	2
Poor	1

In the example above fluorescent fittings when evaluated against the cost criteria were given the rating of very good (4) whereas aesthetically they rated poor (1). The rating given is then multiplied by the weighting to give a total cost score of 24 and an aesthetic score of 10. These figures are then collected into an overall total, in this case 64, which can be compared with the other light fittings. The alternative with the highest rating, in this case the fluorescent fittings, is the optimum choice.

Another similar method of evaluation is the SMART system[14]. This is slightly different from the system above in that, as shown in Table 5.4, the allocated weightings must all be a proportion of 1. In addition the rating of each alternative is not based on a pre-determined scale but on the degree to which the alternative satisfies the criteria. This score is given out of one hundred. In the case of the fluorescent fittings therefore it might be judged that in terms of cost it satisfies the criteria 80 out of one hundred whereas in terms of appearance it only satisfies it

10 our of 100. These scores are then multiplied by the weightings and totalled to give an overall rating. Once again the alternative with the highest score is viewed as the best option.

Table 5.4 The SMART methodology

Method	Initial cost	Maintenance	Aesthetics	Energy impact	Construction time	Total
Weighting	0.24	0.12	0.4	0.12	0.12	1.00
Fluorescent fixtures	80/19	80/10	10/40	60/7	60/7	83

Other techniques available for evaluation rely less on quantification and more on subjective judgement. Some rely simply on a voting system[15]. The amount of quantitative evaluation that takes place really depends on the stage of project development. At the pre-brief stage it may be very difficult to analyse quantitatively the alternatives proposed by the value management team and the choice may simply depend on the preference of the client. If on the other hand the alternatives are well structured then these may be analysed using a weighted matrix.

Value management as a system

Hopefully this chapter has shown that central to the technique of value management is function analysis. Outside of this is a method by which function analysis can be effectively carried out. However because value management is fairly new in the UK and because concepts of value vary, there is as yet no one definitive system; merely a choice of alternative components.

Table 5.5 is a summary of what those components are. A value management system will include an alternative for each component and how those components are put together constitutes the value management system. There is no one correct system and companies should choose the components that best suit them. The summary is not exhaustive. Value management is still in the early stages of development and as the understanding of the components and their use and interaction develop, the list of alternative components will also expand.

Below are given some typical examples of how the components are combined to give value management systems.

The American system

The American system (Table 5.6) is based on a 40-hour workshop carried out by an external team at the 35% design stage. The workshop is structured around the job plan. The practice of American value management uses function analysis loosely based on elements and generated alternatives to them.

Table 5.5 Alternative value management components

Components	Alternatives
Function definition	Based on project function Based on space function Based on elemental function
Function evaluation	Lowest cost to perform function
FAST Diagrams	Use Don't use
Allocate cost to function	Yes No
Calculate worth	Yes No
Generation of alternatives	Brainstorming Other creative techniques
Organisation of the study	Job plan
Group approach	External team Design team Mixture of the two
The VM facilitator	Independent In-house
Format of the value management study	40-hour workshop Two day study Other as applicable to the project
Location	Outside work environment Within work environment
The timing of the study	Inception Brief Sketch design Construction stage Combination of above Continuous process
Evaluation of alternatives	Weighted matrix e.g. SMART Other mathematical techniques Voting Subjective evaluation

A case study of value management in the United States

The study

The project was a Department of Defence training building with an estimated cost of $ 2.4 million. The facilitator was a certified value specialist from a company of value engineering consultants and from a civil engineering background. The team was external comprising an architect, a mechanical engineer, a structural engineer and an electrical engineer. These were selected by the facilitator from consulting organisations. The project was at the 35% design stage. It was single storey reinforced concrete and masonry with pile foundations, concrete floors, built up roof on metal decking. It included a monorail and hoist, sound equipment, compressed air equipment, , exhaust systems, fire protection systems, air conditioning and utilities. The total saving achieved was $154,000 out of $535,980 proposed by the value engineering team. The estimated cost of the study was $21,162 therefore giving a 7.3 return on investment. The study was carried out in a 40-hour workshop and produced the following changes to the project.

Table 5.6 The American system of value management

Components	Alternatives
Function definition	Based on elemental function
Function evaluation	Lowest cost to perform function
FAST Diagrams	Use
Allocate cost to function	Yes
Calculate worth	Yes
Generation of alternatives	Other creative techniques (Ad hoc)
Organisation of the study	Job plan
Group approach	External team
The VM facilitator	Independent
Format of the value management study	40-hour workshop
Location	Outside work environment
The timing of the study	Sketch design
Evaluation of alternatives	Weighted matrix

Value management proposals

- Reduce amount of acoustical CMU
- Delete resilient flooring and seal concrete
- Delete suspended ceiling and paint structure
- Delete buffer area
- Retain exterior insulation
- Reduce crane beam span

- Change long span roof joists to K series
- Use steel framing in lieu of double wall
- Use ledger angle in lieu of joists
- Revise control joists in exterior walls
- Eliminate return air ducts
- Add service shut off valves
- Amend air grill detail
- Apply demand and diversity factors. Delete MDP panels.
- Reduce canopy lighting
- Reconfigure roof intercom systems
- Reconfigure parking lot lighting
- Reduce travel lane width
- Modify storm drainage
- Change type of asphalt paving
- Change type of concrete kerb

The British system

There is, as yet, no definitive British system. The method outlined in Table 5.7 is the one used by John Kelly[16] and anecdotal evidence suggests this is gaining acceptance. This system of value management takes place earlier than the American and is usually carried out by the project design team but led by an external value management facilitator. Function analysis is used to understand objectives and generate alternatives that improve the project.

A case study of value management in the UK[17]

The project was refurbishment of a public building and the objective of the value management study was the production of a control brief document. The study and the resultant brief covered various sections as outlined below.

Examination of strategic issues

The strategic issues that were covered by the study included an examination of the users of the building covering both staff and callers. Included was an analysis of the type of caller, their needs and their queuing behaviour. Other strategic issues covered the political background to the development, the community which it serves, the existing building, future potential technology changes, funding, safety and security. The strategic suggestions made by the value management team consisted of:

- Combine building staff with those currently in another smaller building.
- Co-ordinate the maintenance and capital budgets to achieve better value.

Table 5.7 A British value management system

Components	Alternatives
Function definition	Based on project function and/or Based on space function
Function evaluation	Lowest cost to perform function
FAST Diagrams	Use
Allocate cost to function	Yes
Calculate worth	No
Generation of alternatives	Ad hoc creative techniques
Organisation of the study	Job plan
Group approach	Design team
The value management facilitator	Independent
Format of the value management study	Two day study
Location	Outside work environment
The timing of the study	Brief Sketch design
Evaluation of alternatives	Weighted matrix e.g. SMART

An analysis of the importance of time, cost and quality

In this section of the study the value management team was asked by the facilitator where the priorities of the project lay in terms of time, cost and quality. The team concluded that time was the least important and that cost and quality ranked equally.

A function analysis

A FAST diagram was used to assist in the function analysis. The FAST diagram started at the highest level of function definition and summarised that the aim of the project was to reflect a corporate approach. This was followed through on the FAST diagram to the level of provision of spaces of the building such as toilets and baby changing area. Unlike the American style FAST diagrams it did not attempt to show the inter-relationship between spaces and elements. The diagram is useful in that it shows how the spaces in the building relate to each other and the higher level functions.

User and users flow diagrams

The value management team produced flow diagram for public users of the building, security guards, typists, reception areas, interview points, staff generally, finance staff and messenger staff. These diagrams showed how the building would be used by each of the groups and are aimed at improving efficiency of use.

Space definition

This examined the major spaces provided by the project and looked particularly at the quality of the spaces and the environment that they provided. The aim of this was to examine if the quality and the environment was suitable for the building's intended use.

A study of the location and adjacency of spaces

Through use of a matrix and based on the work carried out in the sections above the value management team produced an adjacency matrix which showed the position of all the spaces in the building in relation to each other. Once again this is aimed at maximising efficiency of use by giving to each space a rating of 0 to 5, where 0 represents no adjacency requirement and 5 represents a definite need for adjacency.

A pre-contract programme indicating the main action points

This section was largely self explanatory and was similar to any pre-contract schedule. The main action points indicated were brief to designers, outline proposals, cost plan, bills of quantities and tender period.

Examination of items requiring immediate action

Based on all the exercises outlined above the value management team produced a list of items requiring immediate action. There were 21 items in all including:
- Check the need for a basement fire escape
- Resolve underground drainage
- Establish floor loading
- Carry out security survey

- Reassess size of waiting area and conference room
- Reassess number of reception points, interview rooms and interview booths.

The agenda of the workshop

The study was carried out over two days to the following agenda:

Day one
- Introduction to value management
- Facilitators present information
- Study of strategic issues
- Lunch
- Time, cost and quality study
- Time schedule
- Function analysis

Day two
- Review of the previous day
- User flow diagrams
- Lunch
- Functional space definition
- Quality and environment study
- Adjacency
- Review and close

The value management team

The team comprised the following people:
- Four members of the client body which included three quantity surveyors
- Seven members of staff or building users
- Two architects from the architectural consultants
- The mechanical and electrical engineer

- Two facilitators

The British case study highlights how much value management has developed in the United Kingdom. Although function analysis is still a core part of the study, the workshop also includes a much more global examination of the projects objectives and user needs. In addition unlike the American study there is little emphasis on cost and stress is on improving the use of the building. Unfortunately the example shown above is still quite rare. However as outlined earlier in this chapter value management in Britain is at the cross-roads of its development and it is possible that this type of study will soon become commonplace. After all the above case study was carried out over two days and with the exception of the facilitators used only those already involved in the project. As such costs would be small, yet as the study shows the benefit to the project can be considerable.

Other British case studies can be found in the work by Green and Popper[18].

The Japanese system

Unlike the American and British systems, Japanese value management is not a one-off exercise but a continuous process carried out under the umbrella of the construction project [19]. The Japanese view value management more as a philosophy than a system, that operates at all stages of the construction cycle, including planning, maintenance and environmental protection (see Table 5.8).

A case study of value management in Japan

In Japan the procurement systems for public and private work are significantly different and this naturally has an effect on the way value management is carried out. In the private sector design and build is the usual form of procurement and under these circumstances value management is carried out by the construction company. In the public sector procurement is more akin to the traditional form of procurement where design and construction are separated. In this case value management is carried out in two stages. First by the in-house designers or consultants and based on the design as it progresses, and second by the contractor. The Value management becomes part of the procurement and a typical process is illustrated below. This is taken from the Kobe City Housing Company project; a multiple function hall with an audio-visual hall and housing combined in a 17 storey building with basement. It has a total of 14,261 m^2 floor area and total cost of 41 billion Yen. The value management procedure was as follows:

Table 5.8 A Japanese value management system

Components	Alternatives
Function definition	Based on project function
	Based on space function
Function evaluation	Lowest cost to perform function
FAST Diagrams	Use
Allocate cost to function	No
Calculate worth	No
Generation of alternatives	Other creative techniques
Organisation of the study	Job plan
Group approach	Design team
The value management facilitator	In-house
Format of the value management study	Other as applicable to the project
Location	Within work environment
The timing of the study	Continuous process
Evaluation of alternatives	Subjective evaluation

A case study of value management in Japan

In Japan the procurement systems for public and private work are significantly different and this naturally has an effect on the way value management is carried out. In the private sector design and build is the usual form of procurement and under these circumstances value management is carried out by the construction company. In the public sector procurement is more akin to the traditional form of procurement where design and construction are separated. In this case value management is carried out in two stages. First by the in-house designers or consultants and based on the design as it progresses, and second by the contractor. The Value management becomes part of the procurement and a typical process is illustrated below. This is taken from the Kobe City Housing Company project; a multiple function hall with an audio-visual hall and housing combined in a 17 storey building with basement. It has a total of 14,261 m^2 floor area and total cost of 41 billion Yen. The value management procedure was as follows:

- *Step 1*

    Designate contractors which can take part in the tender.

- *Step 2*

    Give the contractors the design carried out by the designers, the costing and the procedures for value management.

- *Step 3*

    Contractors make value management proposals based on the existing design.

- *Step 4*

    The value management proposals are examined by the client organisation and classified as good or bad.

- *Step 5*

    The contractors choose some of the proposals which were judged good and make a final proposal to the client which is submitted with a total cost.

- *Step 6*

    The contractor is selected based on the lowest price which must be the same as, or below, the original price at the time of contractor selection.

Some typical proposals from the above project were as follows:

- Change the frame above the fifth floor from *in situ* to precast concrete.
- Change the formwork below the fifth floor from timber to metal to facilitate reuse.
- Use recycled hardcore in lieu of new.
- Change the position of the roof ducting.
- Change the pipe layout of the sprinkler system.

The above is by no means a standard value management system, as in Japan there are many methods of implementing value management. In some case but by no means all a percentage of the saving is given to the contracting company which proposes it.

Why are the systems different?

The value management systems have developed separately because the construction industries and business cultures within which they exist are different[20]. A study of these cultures leads to a better understanding of the systems that operate within them. In the context of this book it also shows that when importing construction management techniques from other countries consideration must be given to the effect of business culture. Investigation of these differences provides insight for the organisation wishing to develop bespoke value management systems that suit their own business culture. The influence of business culture on value management cannot be overstated and the following illustrates how it affected the development of value management in the Unites States and in Japan.

Differences in the style of management

In the information gathering process American managers place greatest emphasis on action in the immediate period and are satisfied with fast results. The Japanese manager on the other hand places greatest emphasis on ideas and looks to the future; obtaining most satisfaction from a range of possibilities as opposed to a definite solution. This indicates that in the USA a fixed system of value management that shows a definite output maybe desirable, whereas in Japan a more evolving process maybe more appropriate. This is in fact reflected in the two value management systems; the American model provides a relatively short, definite process that produces a tangible set of results, whereas the Japanese model is continuous and less obvious.

In terms of information evaluation the American manager places emphasis on logic and is most satisfied with solutions. The Japanese manager on the other hand emphasises human interaction at this stage, largely based on past experience. Satisfaction is obtained not from a solution but a more holistic assessment of the problem. In terms of evaluation of ideas produced by value management it means that in the USA a mathematical form of evaluation maybe preferred, whereas in Japan a more intuitive system maybe appropriate.

Differences in management systems

Styles of management lead to differences in management systems[21]. The Japanese and Western management systems are different and this has a direct relationship on the operation of value management. Table 5.9 summarises the main differences in management systems.

Examining the systems of value management in different countries it can be seen how they have been influenced by management systems. With both British and American value management the single most problematic area has undoubtedly been the human relationships and this is largely a result of the management system being dependent on the individual. Unlike in Japan group behaviour is not inherent and is therefore problematic. The need for short term results has also dictated that the American system is very results oriented, with most studies calculating a return index that shows the saving achieved in relation to the cost of the value management study. This invariably leads to value management teams trying to find as greater savings as possible, often at the cost of overlooking function analysis. The emphasis on logic and reason has also resulted in an American approach that is much more definite than the Japanese, which although a clear system is integrated and less well defined. In addition any business culture that is based on long term planning will probably be more receptive to the concept of value as opposed to cost. This is also true of a system controlled by ideology as opposed to one controlled by task.

Table 5.9 Differences in management systems

Japanese management depends on	Western management depends on
The group[21]	The individual
Long term orientation	Short term results
Ambiguity of responsibility	Clearly defined roles and tasks
Control by humanism	Controlled by logic and reason
Employment of people	Employment of function
Harmony	Seen to be fair
The acceptance of unequal relationships[22]	Equal relationships
Constant improvement	Results
Respect for age and seniority	Less important in business relationships
Decision making by consensus	Decisions by seniors
Control by face saving	Control by superiors
Task control	Ideological control

The relationship between value management and quantity surveying

This chapter should have illustrated that by its very nature value management has very little relationship to the traditional role of the quantity surveyor. This is not to say that quantity surveyors are incapable of broadening the scope of their traditional service to encompass value management, they are, and some quantity surveying practices already offer value management services. However development of value management has not been in the quantity surveyors field because the quantity surveyor is better disposed to the role of the value management facilitator than any of the other professions, it is simply because the Royal Institute of Chartered Surveyor (RICS) have made positive efforts to encourage its development.

Conclusion

The concept of evaluating functions as a means of assessing value is problematic for two reasons. First, value is subjective and second value changes with time.

A piece of jewellery such as a wedding ring has fairly limited functions: it may be to show that the wearer is married, or it maybe purely for decorative purposes. The value to the owner though, particularly if they have owned it for many years, maybe much greater than indicated by the function because it has sentimental value. Most of us acknowledge that there is such a thing as sentimental value and most of us recognise that this value, although real, cannot be defined or quantified.

In addition to its subjective nature the value of items may vary over time. As outlined at the beginning of this chapter the value of a battery operated radio may drastically increase when a snowstorm cuts the electricity supply. In a less dramatic example the value of some goods, such as children's football kits, may diminish simply because a newer version is brought onto the market.

We have therefore no real measure of value other than an assessment of the value to an individual or group of individuals at some given point in time. What we do have, however, and what we can always achieve, is a definition and evaluation of the functions that we either need or want an item to perform. These distinction between needs and wants is an important one. In value management a function does not only apply to what the project needs to do but also to what we would like it to do. If only needs were considered all buildings would be square with flat roofs and linoleum floors. Clearly though many clients wish to provide prestige and comfort in their buildings and these can be equally defined as functions.

The big problem with function analysis in the field of construction is that it is difficult to decide who's functions and values to assess. The value to the government who support the construction of a new hospital may be increased political support in that area. The value to the health authority maybe an increased and better facility for their patients and staff. The value to the staff maybe an improved working environment and the value to the local people maybe a better facility in their area that may reduces delays and travelling time. On top of this is the value to the society, which is the provision of better infrastructure that improves the quality of life for all. The question that arises is which of these subjective and changing values should be assessed. For the purpose of value management the answer is all of them.

But is it really possible to make an assessment of the needs and wants of all those involved in a hospital project, from the politicians down to the patients? The answer in the context of value management is that we translate the needs and wants of those with an interest in the building into functions, that is what the users need and want the building to do. This is not an easy exercise and skill is required in teasing out from the multitude of conflicting interest the true functions of the building and then providing the most economical design solution that meets those functions.

Although function analysis is the central pivot of value management other components are required for function analysis to operate effectively. Within these other components there are alternatives and the choice of alternatives which constitute the value management system is dependent on many factors such as the project, the time scale, the design team and the business culture of the industry. It therefore follows that no one system of value management is correct, as the most effective system will vary with the project. Flexibility therefore is a key to good value management.

Understanding of the component parts of value management and the best time to use them is vital for successful value management. However value management is a fairly new technique and information on when to use what is not always available and may not be for many years to come. More practice of value management is needed, along with more research, before these questions can be answered.

One of the biggest mistakes that is made with value management is to confuse it with cost reduction. We hope this chapter has illustrated that value management has in fact very little to do with cost, it is a design process. The fact that cost reduction often comes about as a result of value management is more a consequence of it than an objective.

---

[1] Chambers English Dictionary (1989) 7th Edition W&R Chambers Ltd and Cambridge University Press.

[2] Fong, P.S.W. (1996) VE in construction: a survey of clients' attitudes in Hong Kong. *Proceedings of the Society of American Value Engineer International Conference* Vol. 31.0.

[3] Palmer A.C. (1992) *An investigative analysis of value engineering in the United States construction industry and its relationship to British cost control procedures.* PhD thesis Loughborough University of Technology.

[4] Miles, L. (1967) *Techniques of Value Analysis and Engineering,* McGraw Hill Book Company, New York.

[5] Mudge A. (1971) *Value engineering, a systematic approach,* McGraw Hill Book Company, New York.

[6] Heller E.D. (1971) *Value management, value engineering and cost reduction,* Addison-Wesley, Reading Mass.

[7] Snodgrass T. and Kasi M. (1986) *Function analysis- The stepping stone to good value* Department of engineering professional development, University of Wisconsin.

[8] Bytheway C.W. (1965) Basic function determination technique *Proceedings of the Society of American Value Engineers Conference* April Vol. 2 pp-21-23.

[9] Kelly W. (1986) *You and value what not,* VEST, Walla Walla Washington State, USA.

[10] Ellegant, H. (1990) Value engineering before you draw! A better way to start project. *Seminar notes: The value management of projects,* Heriot Watt University, Edinburgh, UK, 24th April.

[11] Dell'Isola, Al. (1988) *Value engineering in the construction industry,* Smith Hinchman and Grylls, Washington DC.

[12] Barlow D. (1989) *Successful interdisciplinary ad hoc creative teams.* Applessed Associates, Ohio.

[13] Kelly J.R. and Male S.P. *The practice of value management enhancing value or cost cutting*, Department of Building, Engineering and Surveying, Heriot Watt University, Edinburgh, Scotland.

[14] Green S. (1994) Beyond value management: SMART value management for building projects. *International Journal of project management,* **12** (1), pp.49-56.

[15] J.J. Kaufman Associates, 120006 Indian Wells Drive, Houstan, Texas, 77066.

[16] Kelly J. and Male S. (1993) *Value Management in Design and Construction - the Economic Management of Projects.* E&FN Spon, London.

[17] A VM study facilitated by John Kelly, Senior lecturer, Department of building engineering and surveying, Heriot Watt University, Edinburgh, Scottland.

[18] Green S.D. and Popper P.A. (1990) Value engineering The search for unnecessary cost. *The chartered Institute of Building,* Occasional Paper No. 39, May.

[19] Society of Japanese Value Engineers. (1994) *Management of construction in a changing period.* (Translated from Japanese.) Society of Japanese Value Engineers, Tokyo.

[20] Palmer A. (1996) A comparative analysis of value management systems in the UK, USA and Japan. *CIB International conference, Construction education and modernisation* Beijing, on CD-ROM.

[21] Baba, Keiso, (1990) Principal nature of management in Japanese construction industry. *Journal of Construction engineering and management.* **116** (2), pp. 351-363.

[22] Walker A. and Flanagan R. (1991) *Property and construction in Asia Pacific* Blackwell Science, Oxford.

6   Constructability

Introduction

The terms 'constructability' and 'buildability' will not be found in any standard dictionary. They are terms which are specific to the construction industry and have meaning only to those operating within the confines of the industry. It would be fair to say, that although the principles underpinning these terms are gaining more and more acceptance in a number of countries, the use of the words 'constructability' and 'buildability' is not yet commonplace in the vocabulary of many construction industry practitioners.

In the context of this chapter the terms are taken to be synonymous and can be used interchangeably. In the interests of consistency we have opted for 'constructability' in preference to 'buildability', except when referencing or quoting from authors who have opted for the alternative. We have avoided attaching any subtle difference between the terms, ignoring, for example, the folklore that buildability is British and constructability is American or that constructability encompasses wider system boundaries than buildability.

The industry specific nature of constructability makes it unique in comparison to all other concepts covered in this book. Concepts such as total quality management and reengineering straddle a range of industries whereas constructability can make the unusual claim, that it is the only management concept in the past thirty years to have been designed and developed by the construction industry for the construction industry.

Origins

By comparison with other industries the separation of the processes of design and construction is unique to the construction industry. This compartmentalisation of functions has been highlighted over the years in reports such as the Simon Report[1], the Emmerson Report[2] and the Banwell Report[3]. In response to this perceived deficiency, the Construction Industry Research and Information Association (CIRIA)[4] in 1983 focused attention on the concept of 'buildability'. The view taken by CIRIA was that buildability problems existed, "probably because of the comparative isolation of many designers from the practical construction process. The shortcomings as seen by the builders were not the personal shortcomings of particular people, but of the separation of the design and construction functions which has characterised the UK building industry over the last century or so." CIRIA defined buildability as "the extent to which the design of the building facilitates ease of construction, subject to the overall requirements for the completed building". The CIRIA definition focused only on the link between design and construction and implied that factors which are *solely* within the influence or control of the design team are those which have a significant impact on the ease of construction of a project.

About the same time in the United States the Construction Industry Institute (CII) was founded with the specific aim of improving the cost effectiveness, total quality management and international competitiveness of the construction industry in the United States[5]. Constructability was, and still is, a significant component of the CII's research and development work. The CII definition of constructability is wider in scope than the CIRIA approach and defines constructability as 'a system for achieving optimum integration of construction knowledge and experience in planning, engineering, procurement and field operations in the building process and balancing the various project and environmental constraints to achieve overall project objectives[6]'.

More recently in the early 90's in Australia, the Construction Industry Institute, Australia (CIIA) has tailored and developed the CII constructability process to Australian conditions. In doing so the CIIA has amended the CII definition of constructability to: 'A system for achieving optimum integration of construction knowledge in the building process and balancing the various project and environmental constraints to achieve maximisation of project goals and building performance[7]'.

The goals of constructability

The goals of constructability are determined by the scope which constructability is intended to cover. The 1983 CIRIA definition limited the scope of the concept to the relationship between design and construction. This is illustrated in Figure 6.1.

In systems terms the delineation of the scope of a conceptual model is known as the system boundaries. The system boundaries of the CIRIA model are quite narrow, viewing constructability purely as a design oriented activity. Griffith[8] takes the view that this approach is in fact highly restricted and has resulted in the loss of some momentum in the uptake of constructability in the UK. Several researchers have discussed the difficulty in arriving at the appropriate boundaries for the constructability model[8, 9, 10]. On the one hand, if the boundaries are too wide, there is the inherent danger of applying a simplistic approach which equates constructability with a set of motherhood statements which have very little prospect for practical implementation. Conversely, if a very narrowly focused approach is taken then this may fail to realise the full potential of the concept.

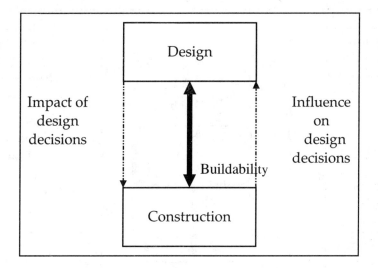

Figure 6.1 The scope of "Buildability" as defined by CIRIA

A workable concept of constructability needs to recognise that there are many factors in a project environment which impact on the design process, the construction process, and the link between design and construction and the maintenance of the building. This can be illustrated diagramatically as shown in Figure 6.2[11].

Figure 6.2 demonstrates that there are many factors in a project environment which impact on the design process, the construction process, and the quality and performance of the finished product. Only when the complex interaction of these factors is acknowledged can the potential of constructability be achieved. Of prime importance is the acceptance of the view that buildability does not equate simply to the ease of construction but is also concerned with the appropriateness of the finished product. This can be seen in one definition which defined buildability as[12] 'the extent to which decisions made through the whole building procurement process, in response to factors influencing the project and other project goals, ultimately facilitate the ease of construction and the quality of the completed project.'

(This definition has been adopted by the New South Wales Government, Construction Policy Steering Committee [13].) Moreover, attention to constructability does not cease with the completion of the building. The constructability of maintenance activities, for example in the installation, removal and replacement of materials, finishes, services and equipment is equally as important throughout the life of the building as the constructability of the initial construction phase.

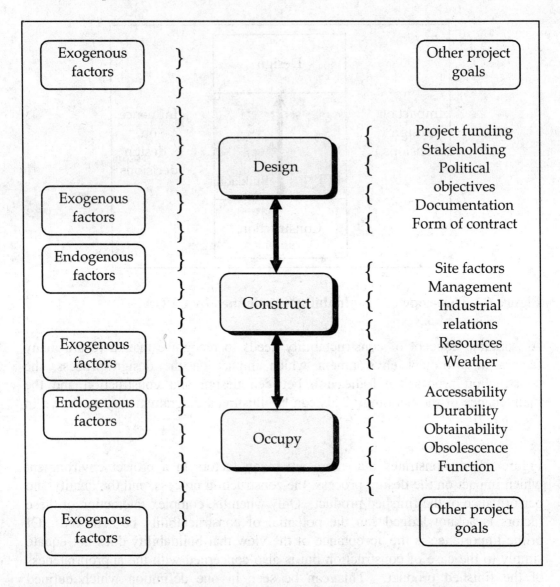

Figure 6.2 performance framework for managing constructability

Decisions which are made upstream of the design stage can impose constraints on the design decision process. At the same time, decisions which may not have been made by the designer concerning intermediate functions between design and construction such as documentation, contractor selection, choice of contract form procedures and so on, may have a significant impact on the construction process. Similarly the maintenance manager who is conventionally regarded as being downstream of the design and construction decision making process will be significantly affected by upstream buildability decisions. Referring to Figure 6.2 it can be seen that each stage of the project, from design through to occupancy, is influenced by forces such as exogenous factors, endogenous factors and project specific goals. (An exogenous factor is one which originates from, or is due to external forces. An endogenous factor is one which originates from, or grows from within.) An approach which takes into account the influence of exogenous and endogenous factors, together with project goals, must, of necessity deal with issues of some complexity. If the complex nature of the procurement process is not acknowledged there is the danger that constructability can lead to common denominator strategies which are of little practical use.

Implementing constructability

Constructability Principles

The previous section stressed the need to take a balanced view and establish appropriate system boundaries if the goals of constructability are to be achieved. This question of balance has been addressed by CII Australia[7] who, in conjunction with the CII[6], has produced a best-practice, how-to- do-it constructability manual (For a detailed explanation of the CIIA Constructability Files in particular and constructability principles in general readers are referred to Griffith A and Sidwell AC)[5]. The CII primer and the CIIA constructability files are not just comprehensive checklists for industry practitioners but are also in keeping with the conceptual model of constructability illustrated in Figure 6.2. The contents of the CIIA Constructability Principles File are as follows:

- Implementation advice on how organisations can establish a constructability program.
- Flowcharting indicating the applicability of the principles of constructability at the various stages of the project lifecycle.
- Executive summaries of the principles of constructability.
- Twelve principles of constructability.
- Database to record examples of savings from constructability.

The CIIA advocates a structured approach which identifies the following five stages in the procurement process:

- feasibility
- conceptual design
- detailed design
- construction
- post construction

The 12 principles are then mapped onto the procurement process. The twelve principles are as follows:

1. Integration - (Constructability) must be made an integral part of the project plan.
2. Construction knowledge - Project planning must actively involve construction knowledge and experience.
3. Team skills - The experience, skills and composition of the project team must be appropriate for the project.
4. Corporate objectives - (Constructability) is enhanced when the project teams gains an understanding of the clients' corporate and project objectives.
5. Available resources - The technology of the design solution must be matched with the skills and resources available.
6. External factors - External factors can affect the cost and/or program of the project.
7. Programme - The overall programme for the project must be realistic, construction sensitive and have the commitment of the project team.
8. Construction methodology - Project design must consider construction methodology.
9. Accessibility - (Constructability) will be enhanced if the construction accessibility is considered in the design and construction stages of the project.
10. Specifications - Project (constructability) is enhanced when construction efficiency is considered in the specification of the development.
11. Construction innovation - The use of innovative techniques during construction will enhance (constructability).
12. Feedback - (Constructability) can be enhanced on similar future projects if a post construction analysis is undertaken by the project team.

Table 6.1 illustrates the distribution of the CIIA 12 principles over the 5 stages of the procurement process. The principles are plotted on a three point scale of importance depending on their location in the procurement stages. For example external factors is of very high importance **[6]** at the feasibility stage, **(6)** is high important at the conceptual design stage but is of lesser importance 6 at the detailed design stage, and has no influence on the construction and post construction stages. These levels are by way of guidelines to users of the CII principles method and are not unchanging. They are simply intended to help the user determine the principles which are likely to be of relevance and of significance at particular stages in the project life cycle.

CONSTRUCTABILITY

Table 6.1 The distribution of the 12 principles over the procurement process

Feasibility	Conceptual design	Detailed design	Construction	Post construction
[1]	[1]	1	1	1
2	[2]	2	2	4
[3]	[3]	[3]	(7)	[12]
[4]	[4]	(5)	(9)	
[5]	(5)	6	[11]	
[6]	(6)	(7)		
7	[7]	[8]		
8	[8]	[9]		
	[9]	[10]		

Note: [ ] very high importance, ( ) high importance, no bracket means is lesser importance.

The various participants in the project process will have different roles and responsibilities in terms of the 12 principles and will have different responsibilities at different stages of the project's life cycle. The decisions taken by the participants needs to be co-ordinated in order to optimise the constructability performance of the project. Otherwise, individual participants may take different strategies towards achieving goals within their sphere of influence thus compromising the overall constructability performance of the project.

Figure 6.3 [14] proposes a constructability implementation planning framework which identifies and co-ordinates the decision roles and responsibilities of individual project participants throughout a project's lifecycle. This enables constructability plans to be developed for individual projects, which allows individual participants to identify, not only their own roles and responsibilities but also allows them to see what other participants are or should be doing at every stage of the project life cycle and therefore allows better integration and co-ordination.

The essence of this approach is that constructability can be enhanced by individual participants exploiting construction knowledge to maximise opportunities and develop best options to meet project objectives in a co-ordinated way and also by adopting collective review processes such as value management.

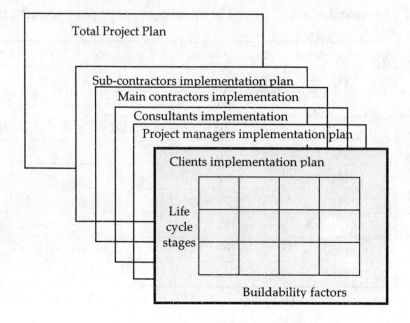

Figure 6.3 Constructability implementation planning framework

Constructability in practice

To successfully implement constructability management, the client or the client's representative should in the first instance put in place a programme which clearly specifies the primary project objectives and allows constructability to be assessed as a project performance attribute. Constructability objectives should also be clearly identified for the different roles and responsibilities of the various members of the project team. This can be done through a performance specification relating to the time-cost, cost and quality criteria addressing the needs of the client and the users, and using the framework (illustrated in Figure 6.3) which co-ordinates the consideration of constructability principles by all the project team members throughout the whole project life-cycle.

Although setting a mechanism in place which co-ordinates the constructability principles amongst the project team members is an important aspect of implementing constructability it is equally important to recognise the significance of the timing of the input by the various team members (Figure 6.4).

The importance of timing is illustrated by the Pareto principle which contends that decisions taken at the early stages in the project life cycle have greater potential to influence the final outcome of the project than those taken in the later stages. 'During the early phases such as conceptual planning and design, the influence of decisions on project cost is high but diminishes rapidly as the project moves into the construction stage. It is vital therefore, to make decisions leading to various design and construction aspect of the project from the first instance. Early contractor involvement in the project is seen to be the desirable approach to attain good buildability'[15].

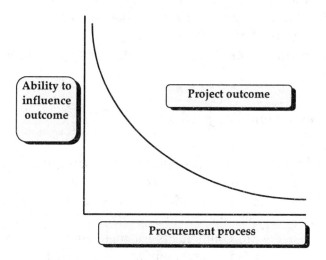

Figure 6.4 Cost influence/ Pareto curve

This is a good juncture to consider the relationship of the design and build form of procurement to constructability (design and construct in Australia). There is no doubt that the design and build form of procurement does provide for early contractor input into the decision making process. Given that CIRIA's initial interest in buildability stemmed from the view that 'buildability problems existed, probably because of the comparative isolation of many designers from the practical construction process'. It may then be thought that the adoption of design and build would go a long way to solving the problem of the dislocation of the process of design from the process of construction. To some extent this is true, and most commentators subscribe to the view that design and build does indeed provide a project environment that is more conducive to 'good' constructability than traditional procurement methods such as selective tendering[5]. However the adoption of design and build, or derivatives such as management contacting and construction management, do not automatically result in a better approach to constructability.

The key to the successful implementation of constructability is in having effective communications between members of the project team, and whilst a design and build form of procurement can streamline the lines of communication, the team members themselves must be committed to the constructability concept if a successful outcome is to be achieved. There is one view[16] that the reason for the occurrence of 'bad' constructability is partly due to changes in emphasis in the education of modern designers and that there is insufficient emphasis given to building construction in the curriculum of architecture courses. An alternative view would be that there is not enough emphasis given to management, or conversely it could be argued that not enough emphasis is given to design appreciation in construction management courses. The permutations are endless, and will always be coloured by the professional background of the commentator. The nub of the issue is that until building professionals receive a multi-disciplinary education then there will always be barriers to achieving 'good' constructability.

Constructability and the building product

The principal goal of constructability is to produce the best product i.e. a building, making the best possible use of resources. Given this focus, it is surprising that so little attention has been given to the relationship between constructability and the building in-use. This is despite the fact that decisions which have a significant impact on the ease of access to components for maintenance and replacement as well as the ease of assembly and disassembly are made in the early stages of the project life cycle. Research work [17] has explored the relationship between decisions taken at the inception stage of the procurement process and the downstream effects on the building in-use. This can be loosely summarised as follows:

Buildings are durable assets which are often threatened with technical and functional obsolescence long before the end of their structural life expectancies. The maintenance and replacement of building components is a common everyday occurrence in the life cycle of a building as a response to physical deterioration, technological obsolescence, changes in performance and functional criteria. Building performance should not therefore be judged at a specific stage in a building's life cycle but should be considered over the life cycle of a building as whole[18]. The potential for a building to extend its useful life is an important factor in its life cycle performance. The operational efficiency of a building through its complete life cycle is determined largely by the characteristics of its original design, the construction or assembly processes and the demands generated by operational requirements, maintenance, alterations and ultimately disassembly or demolition[19]. The level of a building's performance is largely reflected in the quality of decisions taken in the early stages of the project[20].

Efficient building maintenance and renewal is characterised by the need for easy access to, and disassembly and repair or replacement of existing building components. Planning for better performance with respect to these demands should be viewed as being within the domain of constructability. Constructability is often regarded as purely a design issue relating to the ease of initial building assembly. This approach limits and indeed diverts early project decision making away from the consideration of the performance of the building throughout its full life cycle.

The key to improving building performance is in effective information management, particularly at the early project stages where decisions have the greatest potential of influencing project outcomes. The quality of decision making is promoted by the identification of critical issues and the availability of timely and relevant information.

In terms of constructability-oriented maintenance and renewal management, valuable decision support for project policy makers, designers and other stakeholders can be provided by the systematic identification and tracking of important decision situations and planning accordingly to meet required performance objectives. An integrated project decision support framework which can facilitate the implementation of a constructability-oriented maintenance and renewal management strategy is illustrated in Figure 6.5.

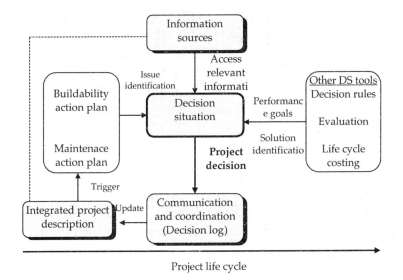

Figure 6.5 Project decision support framework for constructability-oriented Management

Conceptually, the framework integrates the following functions:
- action planning (Constructability-oriented Action Plan)
- cumulative decision recording (Integrated Project Description)
- information access, filtering and processing
- access to relevant decision tools and evaluation techniques

- communication and co-ordination amongst project participants

The constructability-oriented action plan will identify issues having a major impact on constructability and alert relevant decision makers to the issues they need to deal with. The timing of these actions is governed by the development of the integrated project description through each stage of the project life cycle. This project description is the representation of the project in terms of all the decisions which have been made. At the feasibility stage, the project description could be a statement of all perceived project objectives which eventually develops into a design brief. At various stages of the project life cycle, aspects of the project description may be manifested in forms such as sketch design drawings, working drawings, specifications, bills of quantities, and shop drawings. In practice, the compartmentalisation of the different functions in the project process tends to fragment the project descriptions put together by different participants. This contributes to communication and co-ordination problems. The use of computerised tools for integrated applications and networking would allow the development of an integrated project description by which all project participants are informed. The quality of responses to the decision situations generated by the action plan will determine the eventual performance of the project and is enhanced by access to relevant information, and expertise and tools to formulate, evaluate and select optimal solutions. As each decision situation is acted upon, the integrated project description is continually updated, keeping all participants informed on the status of the project and also triggering the progress of the action plan to the following decision situation.

Consideration of maintenance implications at the design stage is not a new idea [21]. However success in this area has been moderate. Maintenance and renewal consideration in the early project stages has been impeded for a variety of reasons, the most critical including: time pressures, unaware decision-makers and ill-informed designers.

Time is critical in the early stages of design and numerous conflicting issues are competing for the attention of the designer. The decision maker invariably has little time to spend on investigating simple yet important operational or servicing issues, let alone more far reaching concepts regarding ease of replacement and renewal considerations. Maintenance management and renewal considerations should be addressed at a policy level by the client or the project manager. Maintenance-oriented design then becomes one of the objectives of the project and maintenance management and renewal becomes an integral part of project management[22]. If maintenance management is established explicitly in the early phases of the project when project objectives are being defined, then resources will be allocated to the examination and investigation of the future maintenance of the building. This implies that particular expertise in the various areas of technical servicing, replacement and component disassembly and reassembly may be retained to advise on the assessment of design alternatives.

Historically the attention to maintenance and renewal performance has been limited because designers responsible for new facilities have been unaware of the post-construction problems and inefficiencies experienced by those who have to maintain and operate them and which their decisions impact upon[19]. Unfortunately designers have limited access to the buildings they design in its post construction period and therefore tend to be divorced from the maintenance problems that flow from poor design.

Decision-makers if aware of possible issues relating to maintenance and renewal rarely have had access to the right information that would aid project decision making. Project decisions made by designers have drawn upon information from three areas: personal knowledge and experience, formal reference sources and project description at that point[23]. In the building industry there has been a tendency for a practitioner to rely on personal knowledge, their own or that of peers[24]. While it has been effective on numerous occasions it has serious limitations[23]. Easy access to formal reference material that is relevant, properly indexed and written in a useable form would provide a reliable base from which designers could make informed decisions.

Decision makers need to be given the required decision support in terms of access to relevant information, expertise and tools. The model illustrated in Figure 6.5 demonstrates that if issues regarding the operation/maintenance/renewal phase are considered then such decision support tools as life cycle costing, maintenance records, post occupancy evaluations and asset registers could be accessed. A systematic approach ensures that the requirements generated by different stages of a project life cycle and the inputs of many compartmentalised project decision processes can be integrated. The overall constructability-oriented management strategy then provides the logical continuity for a single integrated framework that does not overtake the functions of individual project decision makers but serves as a common decision support system to all the project participants. The proposed framework allows the application of information technology (IT) to address the communication, co-ordination and information management problems that confront most project decision making processes.

Ease of access and assembly for building maintenance and renewal are significant issues in the overall performance of buildings over their complete life cycles. A constructability-oriented strategy can provide a logical vehicle for improving maintenance and renewal performance. This is achieved by providing relevant decision support to the early project stages, where the decisions have the greatest impact on overall building performance. The proposed information management framework provides a singular decision support mechanism for all project decision makers and takes into account the dynamic requirements of the project across its complete life cycle. The harnessing of information technology enables the complex co-ordination and communication demands to be met to overcome problems of compartmentalisation.

[Note: The above description is based on the on-going research being undertaken by the Building Performance Research Group at the University of Newcastle into the identification of significant factors in constructability-oriented management.]

Good and bad constructability

The point has now been made repeatedly, that constructability is a dynamic attribute which is project specific. Whilst it is important that examples of constructability, both good and bad, are recorded and a constructability knowledge base is developed and expanded, it is also important to recognise the limitations in terms of the transferability of good constructability features from project to project. In the past, particularly in the UK, many of the recorded case studies seemed to have concentrated on what had gone wrong i.e. examples of bad constructability. Whilst, these may have a salutary effect, greater improvements are likely to be achieved by concentrating on positive aspects. This is why the CII and the CIIA Principles Files[6,7] are an important step forward because by creating a clear conceptual framework the principles underlying examples of good constructability can be identified and *transferred* to new project situations.

The development of a constructability index based on indicators of success which concentrate on what went right in a project rather than what went wrong has been proposed[10]. The underlying assumption is that the greatest gains are likely to be achieved in the management of constructability information and in recording the decision making process during the procurement cycle. This is not to dispute the importance of improvements in construction technology, but to view constructability as a management driven rather than a technologically driven process. In support of this view the Building Research Establishment estimates that 90% of building design errors arise because of failure to apply existing knowledge, strengthening the argument that one of the most important aspects of constructability is not the lack of information but rather the lack of management of information. Of course, gains can, and have been made in the technology areas. Examples, particularly in the UK, include the use of modular co-ordination, design rationalisation, standardisation, pre-casting and dry finishes. However, without the proper tools for managing information for the project decision making process, and without a clear conceptual model to operate from, it is unlikely that the full potential of the constructability approach can be exploited.

As outline above much of the previous efforts to produce a constructability index or scale has been derived from case studies which seem to concentrate on what went wrong with projects. A more productive way forward is for constructability targets to be identified as focal points for decision-making.

Essentially constructability can be considered as a project attribute which is:
- within the influence of those who shape the project process
- is measurable against indicators of success (reflecting the ease of construction and quality of the completed project)

From this working definition, constructability can be represented as having three dimensions. These are :
- the participants (stakeholders and decision makers)
- the constructability factors
- the stages in the building procurement process

Figure 6.6 represents a three-dimensional conceptual model of constructability. The three dimensions define the boundary locations of possible solution spaces where improvements in constructability may be achieved. The three-dimensional model provides a framework to locate the main factors which need to be considered for each project situation to improve constructability.

The participants

The participants comprise two groups. The first group is the stakeholders. These are persons who have some interest in the possible outcomes of the project process. The second group consists of persons who make decisions which influence these outcomes. Some of these possible outcomes are directly or indirectly associated with the constructability of a project.

Constructability factors

The factors (which roughly equate to principles in the CII principles approach) are those which one or more "participants" can influence or respond to in terms of their decisions, and which would impact on the ease of construction or the quality of the completed project. These constructability factors may be exogenous factors, endogenous factors or project goals (other than the goal of constructability (see also Figure 6.2).

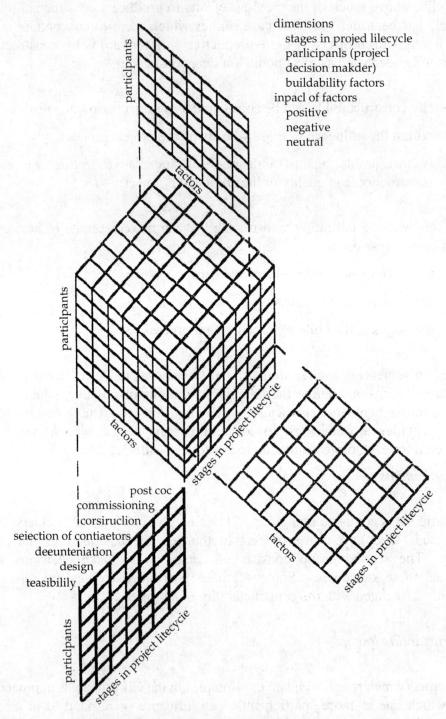

Figure 6.6 A conceptual model of constructability

Stages in the project life cycle

Constructability is not limited to the design-construction subsystem. Decisions which impact on constructability include those which are made upstream of the design stage of the project. At the same time, the standards by which constructability is to be assessed includes the performance of the project after construction.

Application

The approach recognises that in some cases, a factor which is positive in terms of impact on constructability in one project may be neutral or even negative impact on another project. Within imposed constraints, project decisions may be taken so as to minimise or neutralise negative effects, and maximise positive effects. (This work is still at a developmental stage and is linked to research of a more general nature on decision support systems[25].)

Achieving the benefits of constructability

The benefits of constructability, can occur at all stages of the procurement process, although the Pareto principle dictates that the earlier in the process that constructability thinking is incorporated, then the greater will be the impact and the greater will be the potential for time and cost savings and quality improvements. One claim [12] is that the implementation of constructability management can lead to significant quantifiable improvements in project performance in terms of time, cost and quality. In addition to the quantifiable measures, constructability management can also lead to qualitative improvements in the project process as well as the building product. They cite benefits such as:

- better project teamwork
- improved industrial relations
- better forward training
- higher productivity and smoother site operations

A major study in 1988 which investigated the application of constructability in a project management concluded that[15]:

- The contribution of construction personnel to the design of the projects was significant.
- The iterative relationship between construction and design in various project phases led to tangible benefits ranging from cost and time savings and ease of construction, to the elimination of union demarcation and other industrial relations problems.

- The rationalisation of the design, which involves simplification, modularisation and repetition of design detailing, is essential to the achievement of constructability.

- The achievement of constructability is influenced by technical factors such as building technology/ systems, project planning and scheduling etc. in the building process.

- There are many other factors, particularly non-technical factors, associated with the management of building projects (such as project delivery systems, communication, quality of management) which need to be considered as part of the process of achieving constructability.

The CII[6], CIIA[7], and others [5,16,26] have all documented case studies and detailed design situations which demonstrate the benefits of constructability. For readers who wish to explore constructability at detailed level we would recommend Adams[16] and Ferguson[26] who give numerous examples of design detailing and site organisation situations where constructability principles have been applied. This former work lists sixteen design principles and gives examples covering all of these. These design principles are as follows:

1. Investigate thoroughly
2. Consider access at the design stage
3. Consider storage at the design stage
4. Design for minimum time below ground
5. Design for early enclosure
6. Use suitable materials
7. Design for the skills available
8. Design for simple assembly
9. Plan for maximum repetition/ standardisation
10. Maximise the use of plant
11. Allow for sensible tolerances
12. Allow a practical sequence of operations
13. Avoid return visits by trades
14. Plan to avoid damage to work by subsequent operations
15. Design for safe construction
16. Communicate clearly

Other sources give a big picture of the benefits of constructability. Case studies documented by the CIIA [27] typically demonstrate the performance achieved through the systematic application of constructability management. Two examples of these, the Toyota Car Manufacturing Facility at Altona, Victoria, and the Australis Media Centre at the Technology Park in South Australia, give a flavour of the benefits of applying constructability in the following situations:

The Toyota Car Manufacturing Facility comprises a building area of 120,000 square metres and was completed in 107 weeks (3 weeks ahead of schedule), at a cost of AU$161.2 million (AU$18.8 million below budget) to a high standard of quality and safety. One of the significant actions taken in the project was to set up a major roofing and cladding prototype to test and establish the joint and opening details to achieve the required air tightness and weather performance required for the paint shop facilities.

The Australia Media Centre required the construction of a new 7,300 m^2 facility and was completed on a very tight schedule of 9 months at a cost of A$12 million, with cost savings which allowed additional scope of work within the budget. The constructability management strategy employed ensured close co-ordination and liaison between the project team members and early resolution of design, detailing and construction method considerations.

Quantifying the benefits of constructability

Measuring and quantifying the benefits of constructability is not a straight forward matter, particularly if constructability is applied in the broad sense discussed in this chapter. There are several reasons why this is so. In the first instance many of the benefits of constructability, such as better teamwork are qualitative rather than quantitative. In the second instance there is often a synergistic, or knock-on effect in terms of constructability actions. In other words, the whole is always greater than the sum of the parts, thus simply measuring the time, cost and quality improvements of individual aspects of constructability and summating these will not guarantee that the overall impact of constructability has been captured. By way of example if representatives of a local planning authority are included in the project team (as was the case in the Australis Media Case Study[27]) then this may result in fast tracking the approval process, which might in turn have very fundamental implications for downstream activities. The third reason why the benefits of constructability are difficult to quantify, is a methodological one. The ideal way to measure the benefits of constructability would be to conduct parallel case studies of identical projects comparing 'constructability projects' with 'non-constructability projects'. This approach is clearly highly improbable and impracticable and, at best, could only be applied to a very limited sample. It is possible to find instances where very senior executives of major companies are prepared to declare publicly that "they have trialed constructability on three major projects and have saved 5% on cost and 13% on time[27]'. Whilst not disputing the veracity of these claims, the methodology used in quantification is open to challenge on the basis that the comparison relies on a *hypothetical* comparison which assumes an alternative situation where constructability was not used at all, or was used badly.

There have been some attempt to relate managerial actions and building project performance in terms of time, cost and quality[15,28,29,30,31,32]:

"....the practice of constructability analysis is more common in Australia than in the United States or the United Kingdom .... an average of 18 man-months is spent on buildings of 20-30 storeys in height, compared with an average of four man-months on similar US projects, as part of the design process."

A further work that appears to be the first attempt to seriously quantify constructability concludes that: 'The construction time of major projects varies by a factor of almost four to one; The cost implications of the time variation are that savings in holding charges and gain in rental for the quickly constructed projects, compared with the industry average time, can be as large as 50% of the building cost.'

Although it is clear from such studies and also from experience in the United States, that the order of magnitude of the benefits of constructability can be significant, a method of quantifying the indicators of success is not yet a reality.

Conclusion

It may be thought that the virtues of constructability are self-evident and that the principles of constructability are indistinguishable from the principles of good multi-disciplinary team working. This is a reasonable assumption, and one which is difficult to dispute. Constructability is about managing the deployment of resources to their optimum effect. To do so means establishing seamless communication between members of the team. This, in turn, means breaking down of traditional barriers and altering professional mindsets. Builders must be empathetic to the views of architects and vice versa. Clients must be prepared to play their part in responsible decision making. All members of the project team must be prepared to play a pro-active role and address the complete building cycle from inception through to occupation. Expressed in this light we can begin to see why constructability is not yet a commonplace construction management tool. A saying by Jortberg[5] ruefully describes the current situation as 'designers and engineers don't know what they don't know'. Constructability has come a long way from the early CIRIA definition that it is 'the extent to which the design of a building facilitates ease of construction, subject to the overall requirements of the completed building', but it still has a long way to go to attain the goals of the CIIA definition of constructability as being 'A system for achieving optimum integration of construction knowledge in the building process and balancing the various project and environmental constraints to achieve maximisation of project goals and building performance'.

---

[1] Central Council for Works and Buildings. (1944) *The placing and management of building contracts*. HMSO, London.

[2] Emmerson Sir H. (1962) *Survey of problems before the construction industries*: HMSO, London,.

[3] Committee on the placing and management of building contracts. (1964) *Report of the Committee on the placing and management of building contracts*. HMSO, London.

[4] Construction Industry Research and Information Association (CIRIA). (1983) *Buildability: An assessment*. Special Publication 26. CIRIA Publications, London.

[5] Griffith A. and Sidwell AC. (1995) *Constructability in building and engineering projects*. MacMillan Press, London.

[6] Construction Industry Institute (CII). (1986) *Constructability: a primer*. CII University of Texas, Austin, Publication 3-1.

[7] Construction Industry Institute Australia (CIIA). (1992) *Constructability Principles File*. CIIA University of South Australia, Adelaide.

[8] Griffith A. (1986) Concept of buildability, *Proceedings of the IABSE Workshop 1986: Organisation of the design process*, Zurich, Switzerland, May.

[9] Bishop D. (1985) *Buildability: the criteria for assessment*, CIOB, Ascot, Berks.

[10] McGeorge D., Chen S.E. and Oswald M.J. (1992) The development of a conceptual model of buildability which identifies user satisfaction as a major objective. *Proceedings of CIB International Symposium,* Rotterdam, May.

[11] Chen S.E. and McGeorge D. (1993/94) A systems approach to managing buildability. *Australian Institute of Building Papers*, Vol.5.

[12] Chen S.E., McGeorge D. and Varnam, B.I. (1991) *Report to the Government Architect, New South Wales, Buildability stage 1*. TUNRA, University of Newcastle, New South Wales Australia.

[13] New South Wales Government, Construction Steering Committee. (1993) *Capital project procurement manual*: NSW Govt., Australia.

[14] Chen S.E., McGeorge D., Sidwell A.C. and Francis V.E. (1996) A performance framework for managing buildability. CIB-ASTN-ASO-RILEX Proceedings of the International Symposium: Applications of the Performance concept In Building, 9-12 December, Tel Aviv, Israel.

[15] Hon S.L., Gairns D.A. and Wilson O.D. (1988) Buildability: A review of research and practice. *Australian Institute of Building Papers*. **39** (3).

[16] Adams S. (1989) *Practical Buildability*. Butterworths, London.

[17] Chen S.E., London K.A. and McGeorge D. (1994) Extending buildability decision support to improve building maintenance and renewal performance. In *Strategies and Technologies for Maintenance and Modernisation of Building, CIB W70 Tokyo Symposium, 2 (1191-1198) : International Council for Building Research Studies and Documentation CIB*, Tokyo, Japan.

[18] Powell J.A. and Brandon P.S. (1990/1991) Editorial conjecture concerning building design, quality, cost and profit, *Quality and Profit in Building Design*, E & FN Spon, 3-27...Quoted in Mathur K. and McGeorge D. Towards the achievement of total building quality in the building process. *Australian Institute of Building Papers*, Vol. 4.

[19] Bromilow F.J. (1982) Recent research and development in terotechnology: Building maintainability and efficiency research and practice. *CSIRO National Committee of Rationalised Building*, Australia, 3-13.

[20] Speight B.A. (1976) Maintenance in relation to design. *The Chartered Surveyor*, October 1968, quoted in Seeley (1976).

[21] Seeley I.H. (1976) *Building Maintenance.*, Macmillan Press, London.

[22] Kooren J. (1987) *Where Does Maintenance Management Begin?* Building Maintenance Economics and Management. 245-256, E&FN Spon, London.

[23] Leslie H. (1982) *An information and decision support system for the Australian building industry*. CSIRO National Committee of Rationalised Building, Australia.

[24] Mackinder M. and Marvin M. (1982) *Design Decision in architectural practice: the roles of information, experience and other influences during the design process. Research Paper No.19,* Institute for Advanced Architectural Studies, University of York, quoted in Leslie.

[25] McGeorge D., Chen S.E. and Ostwald M.J. (1995) The application of computer generated graphics in organisational decision support systems operating in a real time mode, *Proceedings of COBRA 1995 RICS Construction and Building Research Conference*, Edinburgh.

[26] Ferguson I. (1989) *Buildability in practice*. Mitchell, London.

[27] Sidwell A.C. and Mehrtens V.M. (1996) *Case studies in constructability implementation*. Construction Industry Institute Australia, Research Report 3.

[28] Ireland V. (1983) *The role of managerial actions in the cost, time and quality performance of high-rise commercial building projects*. Unpublished Ph.D. Thesis, University of Sydney, Australia.

[29] Ireland V. (1984) *Managing the building process*. Report: New South Wales Institute of Technology, Sydney.

[30] Ireland V. (1985) The role of managerial actions in the cost, time and quality performance of high-rise commercial building projects. *Construction Management and Economics*, **3**, 59-87.

[31] Ireland V. (1986) *An investigation of US building performance*. Report: New South Wales Institute of Technology, Sydney.

[32] Sidwell A.C. and Ireland V. (1987) An international comparison of construction management. *Australian Institute of Building Papers*, **2**, 3-11.

7   Total quality management

Introduction

In a recent article in the *Times*[1] newspaper a house-owner claimed that after moving into a newly purchased house it took the builder, a national contractor, 103 days to finish the kitchen. In addition, in the short time they had lived in the house they found 112 faults. The couple paid £82, 000 for the property but despite this the house was not ready on the day of completion of the purchase and the painters were still in the house on the day they moved in. The large number of faults that appeared meant that for five months workman from the contractor visited the house three or four times a week. The house owner admitted in the article that the contractor had done all they could to put the defects right. Nevertheless the problems kept appearing.

Naturally the above is a fairly extreme example and in isolation probably not a major problem. However what was really remarkable about the article was the advice of the House Builders Federation to potential purchasers of new properties. This included:

- Buy through a solicitor or surveyor.
- Take care with your contracts as recourse will depend on its terms and conditions.
- As well as thoroughly reading the contract buyers should, with the aid of a solicitor, help to write the contract.

The National House Building Council which, according to the article, is the standard setting and regulatory body of house building also offered their advice:

- Check the builder is registered with the NHBC.
- Talk to previous customers to check the builders standards.
- Check the site is clean and well-managed.
- Turn up on site at an unexpected time.
- A good sign is a fluttering flag which says that the site manager has won an award under the NHBC pride in the job scheme.

- Outside the house look for problems with brickwork, roofing, paintwork, pipework, pipes and drains. A further list is included for inside of the property.
- When there are defects with a property the first port of call is the builder provided he is still in business.

This advice is, in the authors' view, indicative of the endemic problem of poor quality in the construction industry. It is remarkable that these two national bodies, supposedly there to protect the interests of house purchasers, really expect that the average member of the public is able to distinguish a clean and well-managed construction site from a dirty and poorly managed one. It is our guess that to the average member of the public all construction sites are dirty. Furthermore why does the potential purchaser need to buy a house through a surveyor or solicitor? The national contractor mentioned in the article has been building and selling houses for decades. Why can't they do a decent job of selling homes themselves?

Suppose for a moment that the house in question were a car. Imagine if you bought a car for tens of thousands of pounds. Would the manufacturer allow you to drive it away when the paint was still wet? When you buy a new car do you need a solicitor to help you interpret and redraft the contract? Do you need to turn up unexpectedly at the factory to check it is clean and well-managed? Would you be able to distinguish a clean car factory from a dirty one? And how about if it turned out that the car had 112 defects, what would you do? Would you allow the manufacturer to come to your house three or four times a week for five months to fix it? Highly unlikely. More likely you would eventually demand a new car and in all probability the manufacturer, realising the problems were unacceptable, would meet your request.

The problem that the construction industry has is one of poor quality culture. This is reflected by the advice of the two national bodies. The whole industry appears to start from the standpoint that the customer has to look after himself or herself and that it is not necessarily the job of the contractor (or the consultants) to do that.

There can be no doubt that in terms of quality the construction industry has improved enormously in recent years but this improvement appears only to be in regards to certain aspects of the construction process. Sites may be better managed and houses may be better designed than they were ten years ago but final delivery is still not good enough. This is because the industry is not truly customer focused. The industry does not see quality as a whole issue driven by satisfying customer need but as a series of procedures dealing with design, materials or site safety. How *can* a national building contractor with any degree of customer focus allow a buyer spending £82,000 to move into a house that is in the process of being painted?

It is not only the construction industry that is guilty of this lack of customer focus. Manufacturing organisations as well, under competition from Japanese products, have paid the price of poor quality in terms of lost market share. Many of these organisations have now recognised that true quality comes not from a series of checks on products and standards but from a holistic approach to quality that is customer focused. Organisations in the service sector have been particularly successful. National supermarkets are a good example of a holistic approach to quality, with some stores now even offering Sunday school for the children while parents shop.

This then is the essence of total quality management (TQM). It is a customer focused total approach to quality that involves all aspects of, and all people involved in, the organisation.

Definition of TQM

Of all of the techniques outlined in this book TQM is the broadest and most wide ranging. In some ways it is an umbrella under which all other concepts are encompassed. It is possibly for this reason that there is not one accepted definition of TQM. Rampsey and Roberts[2] for example define it as:

'A people focused management system that aims at continual increase in customer satisfaction at continually lower real cost. TQM is a total system approach (not a separate area or program), and an integral part of high level strategy. It works horizontally across function and department, involving all employees, top to bottom, and extends backwards and forwards to include the supply chain and the customer chain.'

Compare this with the following definition[3]:

'TQM is the integration of all functions and processes within an organisation in order to achieve continuous improvement of the quality of goods and services. The goal is customer satisfaction.'

These definitions are in fact quite similar, in that both encompass the fundamental principles of TQM. First, TQM must be a total approach to quality. Whereas in the past quality was concerned with parts of the organisation, such as the final product or customer relations, TQM is concerned with the whole system as an integrated unit. Second, TQM is ongoing. Whereas in the past quality was viewed as a system which could be put in place to improve certain sections of the product or organisation, TQM is a continuous process. The view is now taken that however good a system has become, it can always be improved further. Finally the goal of TQM is customer satisfaction. In the past quality systems have been aimed at improving products and not at customer satisfaction. These are not necessarily the same thing, since it is possible to improve a product without realising that it is not in fact what the customer wants.

Figure 7.1 summarises the three drivers of TQM; integration, customer focus and continuous improvement.

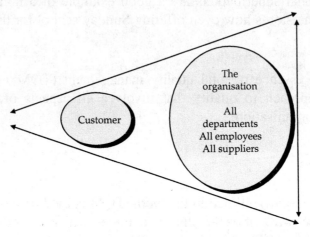

Figure 7.1 Three drivers of TQM: integration, customer focus and continuous improvement

Before examining these three drivers in detail it is worth considering some underlying ideas that support them. First is an examination of what quality is. Second is the historical development of TQM and third is the idea of organisational culture.

What is quality?

As in the case of TQM there are many different definitions of quality[3].

Manufacturing based definitions view quality as the ability to conform to requirements or specification. This measure of quality is objective, in that it is based purely on the ability of the product or service to meet a pre-defined specification or standard. We might, for example, measure if an electric fire produces the correct output of heat or whether a percentage of construction projects were completed on time. The problem with this type of quality definition is that there is no indication that what is measured is in fact what the customer wants. It is an inward looking measure of quality that could not be defined as a total quality approach.

Product based definitions of quality are also objective in that they are based on a measure of a specific characteristic of a product such as, for example, durability or maintenance. A UPVC window could be said to be better quality than a timber window because it lasts longer and needs less maintenance.

User based quality definitions on the other hand, are subjective and evaluate quality based on the extent to which a product satisfies the user. This concept of user satisfaction is closely allied with the idea of function analysis examined in the chapter on value management. User quality can be described as that quality which satisfies the user. In terms of construction this may mean that an item of lower product quality may have higher user quality. In a warehouse building for example the UPVC window mentioned above may have higher product quality. However if the building is low specification, only designed to last twenty years, then timber windows may well have higher user quality because they better suit the needs of the user to provide windows that will last twenty years.

Value based definitions of quality tend to include one of the measures of quality stated above but in the context of cost. A value based examination of the quality of UPVC and timber windows would therefore recognise that the UPVC products are more durable and easier to maintain but this would be viewed in the context of their higher cost. This 'best buy' approach to quality is the one often used by consumer magazines.

So for the modern construction organisation which of the above definitions of quality should be adopted? The answer is that all of them have a role to play. A good construction project will conform to specification and satisfy the user with given levels of quality of the attributes required at the desired price. This may sound rather abstract but these diverse concepts of quality be translated into a more objective list of principal quality dimensions[4].

Performance

This is the primary reason for having the project along with the main characteristics it must have. In terms of a hospital this may therefore be the provision of wards, waiting rooms and operating theatres.

Reliability

This asks if the building will operate for a reasonable period of time without failure.

Conformance

This is the degree to which specification is met.

Durability

This is the length of time a building lasts before it needs to be replaced.

Serviceability

This is the service given after the building is completed particularly with regard to repair.

Aesthetics

This is how the building looks and feels.

Perceived quality

This is the subjective judgement of quality that results from image.

Table 7.1 examines these quality dimensions further. How well does an average building measure up?

Table 7.1 Principal quality dimensions

Quality dimension	Building performance
Performance	Do the majority of buildings achieve their main purpose?
Reliability	Are they reliable?
Conformance	Do they conform to the specification?
Durability	Do they last a longer or shorter period than is required?
Serviceability	Are they repaired quickly and with a quality service?
Aesthetics	Are they aesthetically pleasing internally and externally?
Perceived quality	Does the user and client feel it is a quality building?

Quality therefore is not an easy word to define. It exists at different levels, ranging from the degree to which components of the building meet specification, to the degree that the whole building satisfies the customer. It may be judged either in isolation or relative to some other objective measure such as cost. Quality of a building can be evaluated based on a list of seven principal dimensions. The way to achieve high levels of these principal dimensions is through management of all the processes that deliver them. This is the nature of TQM.

Historical development of TQM

The idea of quality is not new and has its origins in inspection-based systems used in manufacturing industries[5]. In order to reduce the number of faulty goods passed onto the customer, products were inspected during the manufacturing process. The products under inspection were compared with a standard and any faulty goods not reaching the standard were weeded out and either scrapped, or repaired and sold as seconds. These types of inspection based quality systems were found to have several disadvantages:

- Where an inspection did not reveal the faulty item, the problem was passed on to the final user or customer.

- Inspection-based systems are expensive because they are based on rectifying faults.

- Inspection-based systems remove responsibility from the workers and place it onto the inspectors.

- Inspection-based systems give no indication of why a product is defective.

For these reasons and also because products were becoming more complex, inspection-based systems were replaced by systems of quality control based on statistical sampling. One of the gurus of these systems was Deming[6]. The main focus of Deming's work was improvement of the product by reduction in the amount of variation in design and manufacturing. To him variation was the chief cause of poor quality. He believed that variation came from two sources: common causes and special causes. Common causes were those that came about as a result of problems in the production process, whereas special causes were a result of a specific individual or batch of material. In construction terms a common cause might include the sealing of baths against glazed tiles, since there is a fault with the actual design itself and not the material or workmanship involved. A special cause on the other hand might include a bricklayer mixing the mortar in too weak a mix.

In order to achieve improvement in quality through reduced variation Deming outlined a 14 point system of management. These points focus on the process, in that Deming believed it is systems and not workers which are the cause of variation. His points include:

1. Create and publish the aims of the company
2. Learn the new philosophy of quality
3. Cease dependence on mass inspection
4. Do not award business based purely on price

5. Constantly improve the system
6. Institute training
7. Institute leadership
8. Drive out fear and create trust
9. Break down barriers between departments
10. Eliminate slogans and targets
11. Eliminate numerical quotas
12. Remove barriers to pride in workmanship
13. Institute self improvement and a programme of training and retraining
14. Take action to accomplish the change.

Deming believed that once a quality system was set in motion it resulted in a quality chain reaction (Figure 7.2). That is as quality improves costs decrease, as do errors and delays. This causes an increase in productivity and an increase in market share brought about by better quality at lower price. This means the company is more competitive and provides more employment.[4]

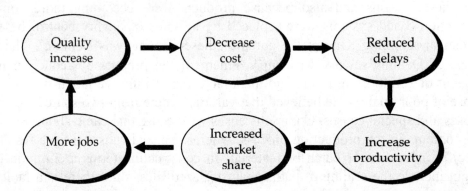

Figure 7.2 Quality chain[6]

Along with Deming, Juran, another American quality consultant, introduced quality control techniques to the Japanese.[4] As the consumption of manufactured goods continued to rise over the next three decades Japanese products became to dominate Western markets. This was because Japanese goods, largely due to their superior quality systems, were of higher quality than their Western counterparts. Whereas quality management had developed as a subject in its own right in Japan, Western quality methods had remained unchanged since their introduction in the 1950s.

There can be little doubt that this gap in quality between American and Japanese goods was one of the drivers of the quality revolution that is taking place in the US and Europe. The word revolution is not used lightly, as recent years have indeed witnessed a fundamental change in the way quality is viewed. Whereas prior to the 1980s quality was internally focused, it is now customer or externally driven. The concept of quality is now strongly related to the idea that an organisation is a series of processes and that for each of these processes there are customers. These customers may be internal to the organisation but by satisfying them the end product will be improved and the end user satisfied. In addition, and as seen from the definitions given above, TQM is now viewed as a much broader philosophy than the traditional techniques of quality control. It now encompass all of the organisation from senior management right through to customers and suppliers. Finally, whereas the older techniques of quality control relied on inspection, TQM relies on prevention[6].

It would be wrong to assume that Deming was the sole guru of quality; other practitioners have disagreed with his approach. Juran's idea of quality for example was not as wide ranging as Deming's and rather than attempting radical change, he sought to improve quality by working within the existing system of the organisation. Crosby's quality philosophy was different again; placing greater emphasis on behavioural aspects than on the statistical analysis used by Deming[4].

The ideas put forward in this chapter therefore are not based solely on the ideas of Deming but on a combination of ideas that are understandable in the context of the construction industry. However this work does not suggest a rigid procedure for TQM because no such thing exits. TQM is a philosophy not a technique.

A philosophy implies a way of thinking and in this context TQM offers its biggest challenge. The way that people think is determined by their culture and in order to instil TQM in the company it is necessary for the organisation to undergo a cultural change. This idea of a cultural shift is closely allied to a modern idea in construction management: that of the paradigm shift.

The need for a paradigm shift

When an industry is developing it is working within the boundaries of its existing system. Any changes that are made, or any developments that take place, do so within those boundaries. These boundaries are useful in that they illustrate the accepted opinion of the discipline and the rules within which it operates. However at some point in history these boundaries become too restrictive and it is necessary to shift the boundaries in order to either explain the changes that are taking place or to facilitate the development that is needed. This shifting of the boundaries is known as the paradigm shift. Many writers on TQM and on the other techniques in this book argue that the construction industry is at the point where the paradigm shift is taking place and that a new paradigm is emerging to replace the old one. It is recognised that the existing paradigm is no longer adequate to meet the needs of a competitive and global construction industry and that it must change. Table 7.2 illustrates this idea of the new and old paradigm[7]. At the beginning of this chapter customer focus was outlined as one of the key drivers of TQM and the table shows paradigms of customer focus broken into the topics of quality, measurement, position, stakeholding and product design.

Table 7.2 The new and old paradigms

Topics	Old Paradigm	New Paradigm
Quality	Meeting specification	Customer value
Measurement	Internal measure of efficiency	Linked to customer value
Positioning	Competition	Customer segments
Key stakeholder	Stockholder	Customer
Product design	Internal sell what we build	External build what the customer wants

An organisation operating under the old paradigm would therefore be a follower, building products it knows will sell. Its main priority would be profit for its shareholders and senior management and it would judge its success against the success of other organisations. It would measure efficiency against predetermined figures with little investigation into what causes the figures and it would control quality through inspection. Finally it would regard quality, in the main, as the meeting of specification. An organisation operating within the new paradigm on the other hand is one that finds out what the customer wants and builds products that meet those needs. Quality and measures of efficiency are all linked to customer value, and positioning in the market is geared towards market share. Finally, short term profits are not the aim as it is the customers, not the shareholders, who are seen as the key stakeholders of the organisation.

In the introduction of this chapter it was outlined that TQM had three drivers: integration, customer focus and continuous improvement. Table 7.2 deals only with one aspect of the new paradigm, namely customer focus. In addition new paradigms are emerging in terms of integration and continuous improvement. In terms of integration barriers between organisational functions are breaking down and human and behavioural aspects of the organisation are more becoming important. Tall hierarchical structures of the past are being replaced with flatter structures with an emphasis on teamwork. With regard to continuous improvement the move is away from top down management operating in a short term perspective, towards more open systems of management concentrating on long term plans that are customer driven[7].

A change in the culture of the construction industry

Paradigms refer to the ways of thinking and acting that is characteristic of a body of knowledge. Culture refers to the social structures that underpin those thoughts and actions. TQM requires a paradigm shift in order to be successfully implemented. For that to happen there needs to be a cultural shift in the organisations that make up the construction industry. This cultural shift that it requires can be summaries in Table 7.3[8].

Table 7.3 The cultural shift required for TQM

From	To
Meeting specification	Continuous improvement
Complete on time	Satisfy customer
Focus on final product	Focus on process
Short term view	Long term view
Inspection based quality	Prevention based
People as cost burdens	People as assets
Minimum cost suppliers	Quality suppliers
Compartmentalised organisation	Integration
Top down management	Employee participation

Organisational culture however is a reflection of social culture and this means that it is difficult to change, since it is vested in the rules that hold society together. However it is not impossible to introduce new organisational culture into the work environment. Evidence of this can be seen in the success of Japanese companies operating in the UK who have successfully planted their organisational culture, albeit with variation, into a social system that wholly different from their own [9].

A Japanese company opening a new plant with its own organisational culture however may be easier than a large British construction company changing from one organisational culture to another. However by realising that a change in organisational culture *is* possible and that it is the only way to achieve TQM can be a greater motivator for change. In addition there are ways in which organisational culture can be changed. Change can be achieved through[7]:

- Management roles
- Training and developments
- Processes and systems
- Employee involvement

Looking at one of these aspects, management roles, in more detail, the traditional view of management's role was one

- Planning
- Organising
- Commanding
- Co-ordinating
- Controlling

However this now needs to include:
- Figurehead
- Leader
- Liaison
- Monitor
- Disseminator
- Spokesman
- Entrepreneur
- Disturbance handler
- Resource alligator
- Negotiator

This is a broader role, which is integrated into the organisation, is more people focused and is a much more realistic assessment of the modern management role. Organisations who wish to adopt TQM must therefore recognise that not only is a cultural change required but also that changing of culture is not the impossible task it is often assumed to be. A proactive approach to cultural change is outlined in Table 7.4[10] which also shows and example of how such a strategy might operate in construction industry.

The first section of this chapter had examined some of the ideas that underpin total quality management, not least the concept of quality and the need for a cultural shift within the construction industry. The next section concentrates on the three main drivers of TQM, namely customer focus, integration and continuous improvement.

Table 7.4 A proactive approach to cultural change

Strategy	Example
Examine the present	Poor sub-contractor relationships
Identify the desired value system	Good sub-contractor relationships
Develop policies to embody value system	Use a smaller group of sub-contractors continuously
	Set up internal mediation procedures to deal with sub-contractors grievances
	Do not hire sub-contractors without first inspecting work that they have done
	Adopt a fair working charter
The total quality strategy	Compile group of sub-contractors
	Devise procedures for mediation
	Draw up fair work charter
Implement total quality management	Communicate new policy to sub-contractors
	Train employees in new methods and policy

Customer focus

The idea of customer focus is important because it gives managers a means of providing products and services that meet customer need. The problem with construction is that it is often difficult to define who the customer is. From the point of view of TQM there are two types of customer: the end user and the internal customer. However in construction even the end user may not be easy to define, since there are often many end users each with conflicting needs. The technique of value management however is geared towards optimising the needs of all the end users of the project. Although this is a new idea, the authors recommend that value management be incorporated into a TQM programme, as in our view, this technique above all others is capable of the complex process of rationalising the conflicting interest that are involved in the construction project.

The idea of the internal customer is one that has been examined before, particularly in the chapter on benchmarking. The idea of the internal customer relates strongly to the idea of the business process. TQM, like benchmarking and re-engineering, is process oriented. As such, the internal customer is the customer of the process or the next person down the chain of production. How to improve the service offered to the internal customer is examined in the next section on integration.

Integration

Traditionally organisations have been separated into departments that carry out a set function, such as estimating. They have also, within these functions, been arranged hierarchically. The need to divide companies this way is accepted practice despite the fact that it leads to some of the following problems[10]:

- Empire building
- Inbreeding
- Diversification of values
- Competition between departments
- Conflicting priorities.
- Different expectations
- Rigid structures

An organisation that is integrated, on the other hand, has a single objective and a common culture. Communications are improved and there is respect for the individual and not the department they work in. One of the main drivers of TQM aims is to break down the compartmentalisation of organisations and move towards more integrated structures. This can be achieved in several ways.

Process management

As explained above an organisation has internal customers who are the customers of the business process. This concept of the business process is central to the idea of TQM, since it is believed that if all internal customers are satisfied then the product or service offered to external customers or end users will also be improved. On a construction project there are usually several organisations involved and this is often a cause of major managerial problems. However the idea of the business process cuts across these boundaries as it is assumed that all the people, in all the companies, are customers. If the main contractor viewed the sub-contractor as a customer (and vice versa) it is assumed that the construction process would run much more smoothly and that a better service and project would be provided to the client and end users. Despite the commonly held but contrary view of the construction industry this is really no different from manufacturing, where different companies manufacture components that make up the final product. Manufacturing industries however appear to accept to a much greater extent than the construction industry, that the overall quality of the final product is only a reflection of the quality of all the materials and workmanship, that have gone into producing it, regardless of the company.

- Interms of TQM therefore a customer is defined as anyone who has the benefit of the work, activity or actions of another. Customers can be categorised as:
- Internal customers who are customers of processes and are within the organisation. (Sales department)
- External customers who are customers of processes and are outside the organisation (Sub-contractors)
- End users who receive, and pay for, the final product. (The client)

It is assumed that satisfaction of internal and external customers through improvement of processes will lead to a higher standard of final product. The idea behind this is that it is not output which should be quality controlled but processes. This is a reversal of the traditional patterns of Western management which have tended to examine output and be highly results oriented. Western systems are generally top down and tend to be designed by management who implement them and wait for results, against which they measure success or failure. This can be described as unitary thinking: that is decisions based totally on quantitative measures of output. The quantitative measures generally have some financial point of reference.

The effects of unitary thinking[10]

Unitary thinking is geared towards numbers but this is a very poor way of examining the organisation and tends to lead to:

Stagnation

If an organisation constantly uses the same targets to measure its success it will undoubtedly stagnant. The environment within which an organisation exists is constantly changing. It makes no sense therefore for the targets against which success is measured to remain constant.

Undermining of quality

Quality cannot be measured purely quantitatively. Reliance on numerical measures of success therefore will undoubtedly lead to focus moving away from quality and onto achievement of the numerical target.

Shortening timescale

Where a numerical target has to be met there is a tendency to focus on the time scale within which it exists. Measuring success against quarterly reports or interim valuations encourage staff to see projects (and the company at large) as a series of short steps or phases. This type of short termism is detrimental to quality and the long-term health of the organisation.

Inhibition of investment

Concentration on short-terms results and targets discourages long-term investment.

Compartmentalisation activity

This is a common problem with large organisations and occurs when employees identify with their own particular department or activity and not with the organisation as a whole. This leads to infighting and lack of commitment to the company as a whole.

Process thinking

The reverse of unitary thinking is process thinking, which rather than concentrating on outputs and numerical measures of it, concentrates on processes. Behind process thinking is the idea that the cause of success and failure lies in the systems that produce the results and ultimately in the processes that make up the systems. The benefits of process thinking include[10]:

Lengthen timescales

Process improvement by its nature is ongoing. As a result a company that is process oriented has a longer term perspective which tends to promote investment and long term success.

Create a company culture

As outlined earlier at the root of TQM is the need for a cultural shift within the organisation. Concentrating on the process can assist it making this cultural shift as it diverts attention away from output, short term results and compartmentalisation.

The inverted organisation

The inverted organisation can also act as a means of integrating the organisation. Rather than seeing the organisational structure with the manager at the top the inverted organisation views that those who deliver the service as being at the top, as they are from the front line.

Employee involvement

A further way that integration can be achieved in the organisation is through employee involvement. This idea of all company employees being part of the organisation is not simply a utopian ideal there for its own sake. Its purpose, through empowering members of the organisation to make decisions and solve problems, is to act as a means of improving the system. Its purpose is to pull employees into the process and to encourage ownership of it. Employee involvement can be achieved in any of the following ways:

Quality circles

Quality circles bring together those employees involved with a particular process to identify, analyse and solve job related problems that relate to that process. Quality circles are important because they encourage ownership, and as a result improvement, of the process.

Quality circles can be negative or positive[10]. Negative circles are introduced only where there is a particular problem with the process. Positive circles on the other hand are permanent groups. Positive quality circles are viewed as the better of the two since there is often more to be learned from examining the process when it was normal than when there is a problem with it.

Essentials of a quality circle
- People
- Skills
- Time
- Place
- Resource

Stages in a quality circle organisation
- Identification - The process that is either problematic or which is to become the subject of a positive quality circle is identified.

- Evaluation - Any problems or potential problems with the process are identified. A distinction is made between problems that can be rectified and those which are outside the control of the company.

- Solution - An optimum solution for improving the process is arrived at.

- Presentation - The results of the quality circle and the proposed improved system are presented to senior management.

- Implementation - The improved process is implemented. However as process improvement is a continuous process the new process must be monitored and if necessary improved further.

Quality circles are particularly useful in solving problems related to direct labour[3] where the problems dealt with are of a practical nature. There have been problems with their use in more general areas where middle management can see them as a threat to their authority.

Cross functional teams

Cross functional teams also increase integration by operating laterally in the organisation as opposed to following the typical hierarchical structure. The basis of cross functional teams is the co-ordination of organisational functions on the basis of processes that run through them.

Training and development

There cannot be employee involvement without training. TQM training will usually fall into three categories of reinforcement of the quality message, skill training relevant to a particular task and principles of TQM[3]. Critical to this last area is problem solving. Problem solving is a huge area of management and for that reason only a brief outline of the major problem solving techniques is outlined below.

Brainstorming

Brainstorming was outlined briefly in the chapter on value management as a means of generating creative design solutions. Brainstorming can also be used as a means of problem solving. Like its counterpart in value management it has certain rules. Before starting a brainstorming problem solving session there should be no definition of the problem other than in totally neutral terms and suggestions of solutions should not be made. Time for suggested solutions to the problem should be limited and all answers should be noted, however outrageous they may seem. Once again unusual ideas should be encouraged as often these generate effective and workable solution.

Pareto analysis

A Pareto analysis involves calculating both the costs and frequency of problems that occur and putting them in order of magnitude so that the highest cost problems are revealed and singled out for improvement.

Cause and effect diagrams

These were invented by Kaoru Ishikawa and are also referred to as fishbone diagrams or Ishikawa diagrams (Figure 7.3)[10].

In the chapter on benchmarking it was outlined that a process has an input, a process and an output and that on organisation consists of hundreds, or possibly thousands, of inter-related processes. In the fishbone diagrams a cause is an input to the process. An effect on the other hand is the result of the process. When the process is working correctly, or is in control, the effect or output of the process is predictable. When the process is out of control the effect is not predictable. Ishikawa believed that the investigation of cause is most critical. He believed that although something may look like a cause, in reality it is an effect of a previous process. By tracing back alone the fishbone this effect can be found. It can only be found however when it is known what the true effect should have been. For that reason it is equally important to study the process while it is in control as it is when it is out of control.

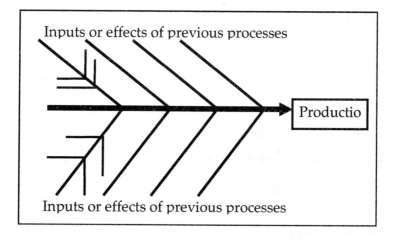

Figure 7.3 Ishikawa Diagrams

Performance appraisal

Performance appraisal is also a method of encouraging employee involvement. However traditional methods of appraisal may not be suitable for a company on track to TQM. Performance appraisal should focus on the objectives of the company and not the individual output.

The one more theory

The one more theory operates on the basis that every job in the organisation should be able to be undertaken by at least two people and that every person in the organisation should be able to do at least one other job in addition to their regular work.

The all-embracing nature of TQM

The all embracing nature of TQM deems it necessary for companies to include both suppliers and customers in their TQM process. Of all the systems covered in this book it is probably TQM more than any other that falls into the category of a soft system. For that reason it is impossible to say what is a typical system of TQM. It is however possible to say what are the characteristics of a company on route to total quality[5].

- Vision
- Attention to customer
- Attack of failure
- Focus on prevention
- Zero functional barriers
- Monitoring of competition
- Senior management actions
- High levels of training

Continuous improvement

The third driver of TQM is that it is a continuous process of improvement. Continuous improvement means that even where an organisation is profitable, with a high percentage of market share, it should still look for ways of improving. It cannot be iterated enough that TQM is not a system but a philosophy, an important part of which is the notion that no system is ever perfect and can always be improved.

Quality costs and the cost of quality

Many people reading this book may by now have concluded their own organisation cannot, for reason of size or profit margins, afford to adopt TQM. However such opinions are often vested in a misunderstanding of what the actual costs of quality are. There are basically three ways of thinking about quality costs and these are outlined below.

Higher quality means higher cost

Increased quality cannot be obtained without an increase in cost and the benefits to the company from increased quality will not compensate for the additional cost involved.

The cost of improving quality is less than the resulting savings

This second view of quality costs is that the additional cost of quality is less than the money lost through re-building or scrapping of work.

The right first time approach

This view of quality cost is much wider than that stated above. Quality costs are viewed as those incurred in excess of those that would have been incurred if the product were built right first time. Costs are not only direct but those resulting from lost customers, lost market share and other hidden costs. Costs are not measured as the costs of re-building or scrapping but the extra over cost that is incurred based what the costs would have been if everything had been built the first time to zero defects. This is the view of most TQM practitioners.

These costs of quality measured against the zero defect can be classified into four categories:

Prevention - This are costs which remove or prevent defects from occurring. Examples are quality planning and training.

Appraisal - These are the costs incurred to identify poor quality products after they occur but before shipment to customers. Inspection costs are an example.

Internal failure - These are costs incurred during the production process and would include scrap and rebuilding costs.

External failure - These are the costs of rejected or returned work and include the hidden costs of customer dissatisfaction such as loss of market share. Most TQM practitioner are of the view that quality costs fall into the iceberg principle, in that only the small portion that is above the surface, such as re-building costs, are visible. The part that cannot be seen because it lies below the surface however is likely to form a much larger proportion of the cost, possibly even as much as 90% of the total.

Universal standards of quality such as ISO 9000

ISO9000 is a set of five world-wide standards that establish requirements for the management of quality. ISO 9000 is not a product standard but a standard for quality management systems. It is used extensively in the European Union. The standard is generic in that it applies to all functions and all industries. By 1992 more than 20,000 companies in Britain had adopted the standard and have become certified and over 20,000 companies from outside the EU have also registered. The Japanese have also mounted a campaign to get their companies registered.

Documentation for ISO 9000 is onerous and requires the writing of quality manuals, the documentation of all relevant procedures and the writing of all relevant work instructions.

- Benefits of certification
- Greater customer loyalty
- Improvements in market share
- Higher stock prices
- Reduced service calls
- Higher prices
- Greater productivity
- Cost reduction

There are also disadvantages of certification. As this chapter has consistently stressed TQM is a philosophy and it is viewed that the rigidity of the ISO 9000 certification is contrary to this ideal. It tends to oppose the very objectives for establishing TQM in that the quality focus moves away from the customer towards the gaining of certification which becomes the primary objective.

Likewise the responsibility for quality moves away from employees and back to the quality department responsible for obtaining certification.

Research in the construction industry has shown that there is no conclusive evidence one way or another to prove the benefit of certification in the construction industry[11]. However it must be borne in mind that quality systems such as ISO 9000 and TQM are not the same thing. In addition the criticisms that have been made of certification may relate not to the systems themselves but to the way they are being implemented.

Change management

One thing that will be obvious from reading this chapter is that the move to become a TQM company will involve change at all levels of the organisation. Change however will not happen effectively by itself; it needs to be managed. Managing change means taking control of the process and shaping the direction that the change will take. The management of change is a four stage process[5]:

- Establish the need for a change
- Gaining and sustaining commitment
- Implementation
- Review

One of the important aspects in the effective management of change is that it must involve all members of the organisation. This is opposed to the current view of where change tends to be decided at senior management levels and then passed down to lower levels of the organisation with little explanation as to its rationale. Alternatively there is the mushroom approach whereby management simply throw some ideas to a situation they do not fully understand and wait and see what happen.

Even when change is managed there are barriers to it. These include cost, lack of time, employee perceptions, industry culture and lack of ability. Recognition and understanding of these barriers to change are an important step to overcoming them.

The methods of TQM

Any book on quality and TQM will include a vast array of quality and TQM methods. Most of these methods have been tested extensively in manufacturing industries but whether or not they are applicable to construction is not known. For this reason they are mentioned below but are not included in detail. Any reader wishing to study them further is referred to the references at the end of this chapter.

- Taguchi methods
- Failure mode and effects analysis
- Statistical process control
- Just-in-time

How to implement TQM

In order to implement and maintain a system of TQM there must first of all be planning at the strategic level. Such a plan should include[10]:

Management of quality

- Quality definitions
- Total quality policy
- Total quality strategy
- Total quality culture

Management of people

- All employees should understand their position and that of others in the organisation.
- Commitment
- Team work
- Education
- Open management

Management of process

- Process design
- Process control
- Process improvement

Management of resources

- Wealth generation
- Cost of quality

- Resource conservation
- Resource planning

Kaizen[12]

One item which needs to be mentioned in relation to TQM is the Japanese concept of *kaizen*. *Kaizen* strategy is probably the single most important concept in Japanese management. The message of the *kaizen* strategy is that not a day should go by without some kind of improvement being made somewhere in the company. In Japan management is perceived of having two functions: maintenance and improvement. Kaizen signifies small improvements made in the status quo as a result of ongoing efforts. It is different from TQM in that it operates within existing cultures and rarely requires cultural shifts. Whether or not *Kaizen* type systems could be made to work in the West is not known.

Current research into TQM in the construction industry

One section of research in the field of TQM looked at certification. Work had examined the impact of the introduction of BS5750/ISO9000 in construction related organisations and highlighted that there are very mixed views on the benefits of certification[11]. Certification was also examined in relation to architectural practices[13], and it was found that in relation to building contractors, architectural practices were slow to adopt Quality Assurance (QA) systems. The percentage of architectural practices with certified QA systems is in fact a very small percentage of the total.

Further work in this field[14] has established that certification may not be the best route to take and that possibly the most effective means of establishing a total quality culture and improving quality, is for the company to develop its own quality improvement team.

Work on quality control among contractors[15] in South Africa showed that contractors recognised the importance of quality systems particularly with regard to their use as criterion for contractor pre-qualification. However in keeping with the work undertaken in UK it was shown that many South African contractors found the formality of quality systems a problem and often failed to realise that it was possible to have quality systems that were based on the needs of their own organisations. Many of the contractors failed to realise that quality system, to be effective, did not necessarily have to be formalised.

Another area which has been examined is the relationship between procurement and quality management[16]. This work is new and although it indicated that contractors may feel that design and build is the most appropriate for addressing quality issues, there is no conclusive evidence to support this.

Conclusion

In a recent lecture on process improvement one of the authors' students voiced the view that if something is not broke why try and fix it. In some ways this idea is the opposite of the TQM culture where it is believed that no system, however good it may appear, cannot be improved further. As this chapter has constantly stressed, TQM, possibly more than any of the other techniques in this book requires a cultural shift for its effective implementation. Without this realisation TQM can almost definitely not be achieved. For this reason this chapter has not tried to give detail of how TQM systems operate but a more general view of the culture and environment required for successful TQM.

TQM is basically an attitude whereby customer focus is the main driver of the organisation. This is achieved within an organisation which is integrated and that is committed to continuous improvement. These are the three basics of TQM and within these a multitude of systems are available and within those systems a multitude of techniques. Some systems of TQM are highly formalised. However there is no evidence to suggest that formalised systems are any better than bespoke systems developed in house.

There can be no doubt that the construction industry is an enormously complex one and that as an industry it does have barriers that appear to prevent the development of a quality culture. Competitive tendering is an obvious example in that contractors who are forced into low prices as a means of securing work will always have to cut corners to stay within budget. Under such circumstances quality is an obvious casualty. Sub-contracting appears another barrier. What ever quality systems a contractor may have in place these can become extremely difficult to implement when the entry of sub-contractors into the industry is so easy and when the means of appointing them takes no account of the sub-contractors own quality. In addition is the difference in the level of management sophistication between the main contractor and the sub-contractor[17]. The skill shortage makes matters worse.

These however tend to be circular arguments of the viscous circle variety. The only way that the circle can be broken is through a cultural shift, not only in the contractors but also in the clients of the industry, the consultants and the sub-contractors. It is the authors' view that this cultural shift has started and that once it gathers momentum the construction industry will change more quickly than has been anticipated.

[1] Kelly, R. (1996) Heartache of a faulty house. *The Times* Wednesday 6 November p. 42.

[2] Rampey J. and Roberts H. (1992) Perspective on total quality. *Proceedings of total quality Forum IV* Cincinnati, Ohio, November. Cincinnati, Ohio.

[3] Vincent K. O. & Joel E. R. (1995) *Principles of total quality* Kogan Page, London.

[4] Evans J. R. and William L. (1993) *The management and control of quality.* West Publishing Company, Minneapolis.

[5] Asher J.M. (1992) *Implementing TQM is small and medium sized companies.* TQM Practitioner series. Technical Communications (Publishing) Ltd. Letchworth, Hertfordshire.

[6] Deming W.E. (1986) *Out of the crisis.* Massachusetts Institute of Technology Cambridge, Mass.

[7] Bounds G., Yorks L., Adams M. and Ranney G. (1994) *Total quality management: Toward the emerging paradigm*, McGraw Hill, New York.

[8] Baden H. R. (1993) *Total quality in construction projects. Achieving profitability with customer satisfaction* Thomas Telford, London.

[9] University of Cambridge (1993-1994) *The Financial impact of Japanese production methods in UK companies* Research papers in management studies, No. 24.

[10] Choppin J. (1991) *Quality through people A Blueprint for proactive total quality management.* IFS Publications, UK.

[11] Hodgkinson R., Jaggar D. M. & Riley M. (1996) Organising for quality, *Construction modernisation and education. CIB Beijing International conference,* Beijing, On CD-ROM.

[12] Imai M. (1986) *Kaizen:The key to Japanís competitive success.* McGraw Hill, New York.

[13] Emmitt S. (1996) Quality assurance-More than a marketing badge, *Construction modernisation and education. CIB Beijing International conference,* Beijing, On CD-ROM.

[14] McCaffer R. and Harvey P. (1996) Total quality construction: a never ending journey. *Construction modernisation and education. CIB Beijing International conference,* Beijing. On CD-ROM.

[15] Rwelamila P., and Smallwood J. (1996) The need for implementation of quality management systems in South African *Construction modernisation and education. CIB Beijing International conference,* Beijing. On CD-ROM.

[16] Rwelamila P. D. (1995) Quality management in the SADC construction industries. *International Journal of Quality and reliability management.* **12** (8), 23-31.

[17] Cheetham D. W. (1996) Are quality management systems possible. *The organisation and management of construction, Vol. 1: Managing the construction enterprise.* CIB W65 International symposium, Glasgow, pp.364-365.

8 Current construction management issues in Western and Chinese construction industries

(Note: A major part of the content in this chapter first appeared in the Proceeding of International Conference on Project Cost Management, Beijing China, 25 - 27 May 2001.)

Introduction

In our view there has been a tendency to assume that because China is in a transitional stage from a planned economy to a market economy that China should accelerate the change-over process by adopting modern Western management concepts. This may not necessarily the case.

It is perhaps worthwhile emphasising the point, that in Western society there is no single unified approach to management in general or construction management in particular. For example, when UK professionals tried to use the US system of value engineering they found that they couldn't make it work. The reasons for this were numerous but largely related to the original objectives of the value engineering studies[1]. The US system of value engineering was born out of a need for greater accountability on government projects. (Almost all value engineering activity in the US is government work.) The situation in the UK was very different. The UK quantity surveying system (cost control system) provided all the accountability that was needed. Value engineering was required to provide a platform for the examination of value as opposed to cost. This is an illustration of how cultural differences between Western countries has lead to different approaches to the same management concept.

Differences also exist between the US approach to reengineering and the European approach. The Europeans generally finding the approach of Hammer and Champy much too aggressive for European acceptance[2]. The head of Siemens, Heirich von Pierer, writing on Business Process Reengineering (BPR) has been quoted to the effect that 'I don't feel completely comfortable with the radical thesis of Mr. Hammer. Our employees are not neutrons, but people. That's why dialogue is important.' Holtham goes on to express the view that BPR needs to be rooted in distinctive European managerial features, in terms of the 'acceptance of the humanistic and holistic stream in European thought, in contrast to the more mechanistic and fragmented US approach with the promotion of the concept of collaboration, between levels in the organisation, across organisations, between supplier and customer, and also across national boundaries. There would appear to be little variance between Holtham's vision of BPR in Europe and Hammer's description of the application of BPR in the United States. Perhaps it is more a disagreement on the method of delivery than the message being delivered. As Holtham concludes, 'The core elements of BPR have value beyond the evangelical North American approach, and BPR has value for Europe, if it is set in a European context'.

Cowan, one of the principal architects of the modern partnering movement stresses that 'Partnering is more than a set of goals and procedures; it is a state, a mind, a philosophy. Partnering represents a commitment of respect, trust, co-operation, and excellence for all stakeholders in both partners' organisations'[3]. Despite the fact that partnering, has been highly successful in the US, Australia and the UK it has not gained acceptance in Japanese construction industry where its general philosophy is deemed to be so much a part of Japanese business culture that partnering as a business process is, in effect, redundant.

Whilst on one hand making the argument that some Western construction management concepts may not be appropriate to the Chinese construction industry culture, we are not trying to make the case that there is no opportunity for cross fertilisation between Western and Chinese cultures. When we were first introduced to the Tangible Construction Market or TCM[1] we were fascinated by

---

[1] In 1993, the Chinese Ministry of Construction (MOC) formally published its planning outline for the construction market, based on reforming the allocation of tenders - controlling the tendering process rather than attempting to control tenderers. The State Council resolution followed joint promulgation in July 1994, by the Ministry of Supervision and the Ministry of Construction, of ministerial directions regarding corruption and unfair competition, with emphasis on:

- Registration systems for construction projects
- The introduction and promotion of public bidding systems
- Banning Government Departments at the various levels from interfering with the bidding system
- Strengthening management to intensify supervision of bidding firms, to prevent collusive bidding
- Educating officials involved in the public bidding process to be honest
- Intensifying supervision of projects, as well as improving investigation and management of cases of corruption or mismanagement

the novel approach which China had taken in tackling a long-standing problem of administrative accountability. The TCM is an interesting phenomenon. It is so named because it is just that – a transparent, auditable process conducted in a tangible – physical – location where all of the officials and construction industry representatives come together to conduct the public bidding process. The word 'tangible' is used to clearly distinguish the new process from previous practices of awarding construction contracts behind closed doors or otherwise in secret – an 'intangible' market. It would appear that the approach taken in establishing the TCM is very much in line with the reengineering concept as advocated by Hammer[4] that "Overtime, corporations have developed elaborate ways to process work. Nobody has ever stepped back and taken a look at the entire system… the imperative for reengineering is to achieve a quantum leap forward rather than small continuous gains". We believe that the TCM has captured the spirit of reengineering by achieving 'discontinuous improvement' and is an example of how Western reengineering principles can cross a cultural divide. Indeed it could be argued that the TCM approach to tendering is a model which might be usefully deployed in any market driven economy.

Cultural trends

Having made the argument that there is no world-view of construction management or construction economics, there does however seem to be a universal trend in terms of greater stakeholder involvement in the procurement process. (By procurement, we mean the complete building cycle from the inception stage of a building project to the completion stage when the building is ready for occupation).

The dominant message from market economies such as the United Kingdom and Australia[5,6] is the key role of the client in activating a cultural shift in the industry through the adoption of modern management concepts. This is summarised by Latham who states that 'implementation begins with clients. Clients are at the core of the process and their needs must be met by industry'. Latham then goes on to recommend that 'Government should commit itself to being a best practice client. It should provide its staff with the training necessary to achieve this and establish benchmarking arrangements to provide pressure for continuing improvements in performance.' This has been expressed more succinctly if more crudely, as ' the client having the power of the cheque book'. The emphasis on greater client involvement is closely parallel in the Chinese construction industry by the 1993 Chinese Ministry of Construction (MOC) planning outline for the construction market (see footnote to TCM).

One consequence of these ministerial directions was the establishment of the TCM.

Generally, it would appear, that Government clients are taking on board the directives of Egan, Latham, Gyles and the MOC and are now committed to continuous improvement through the strengthening of the management of the construction industry. Whilst this initiative is to be applauded there is still however a good deal of ambiguity about how this might be achieved. There are probably several reasons why this is so. One reason may lie in the fact that many, if not all of the current management concepts are philosophically grounded, if not in systems theory, then at least in a holistic approach. We would contend that this common parentage has given rise to difficulties in terms of identifying concepts as individual branches of the same family tree. This lack of differentiation between current concepts is typified in comments such as 'Constructability is not just value engineering or value management' or 'is reengineering replacing total quality? or more confusingly, 'As partnering is to the project, total quality management is to the construction company'.

Current issues

The culture of the construction industry in Western has changed significantly in the last decade. The traditional culture of adversarial relationships is changing and major Government clients have recognised that a commitment to best practice is the way to developing good long term relationships.

Australian State Government Authorities such as the New South Wales Department of Public Works and Services (DPWS) actively promote best practice through the use of contractor pre-qualification schemes using prequalification criteria.

The DPWS contractor accreditation scheme [7] lists the following construction industry best practice initiatives:

1. Commitment to client satisfaction
2. Quality management
3. Occupational health and safety and rehabilitation management
4. Co-operative contracting
5. Workplace reform
6. Management of environmental issues
7. Partnering
8. Benchmarking
9. Another area of best practice nominated by the contractor and accepted by DPWS.

For contracts valued over Aus. $20M contractors must have a demonstrated record of commitment to, or a corporate program for the early implementation of criteria 1 to 6 and at least one of the reform initiatives 7 to 9.

Under the scheme, contractors who achieve best practice accreditation will be offered significantly more tendering opportunities than those contractors who are not accredited. It is anticipated that ultimately, only those contractors accredited under the scheme will be eligible for selection to tender for contracts valued at over Aus. $500,000. (A good example of the power of the cheque book in a market economy.)

The DPWS example is given, not by way of promoting DPWS, but by way of illustrating the central theme of this chapter that clients are taking on board the introduction of modern management concepts and are applying pressure to the construction industry to accelerate their adoption. They are doing so either explicitly by the use, for example, of formal reengineering principles or implicitly, as in the case of China, in the adoption of a radical approach to tendering in the form of the TCM. In the decade of the 80's the practising construction manager, or the construction management undergraduate would not have heard of concepts such as value management; total quality management; constructability; benchmarking; partnering and reengineering. In the 90's there was a greater awareness of these concepts, and clients, particularly government agencies pushed hard for their adoption. However, there is also a good deal of ambiguity about the nature and usefulness of the concepts[8].

As previously stated one reason may lie in the fact that most of these modern management concepts are philosophically grounded, if not in systems theory, then at least in a holistic approach which encourages increasing stakeholder involvement.

For many years critics of the construction industry have dwelt on the perceived problems of fragmentation and compartmentalisation. Many of the ills which have beset the industry have been blamed on the inability of the industry to see the big picture. Many of the advocates of the techniques that 'their' concept rectifies this. For example Hellard[9] advocates that 'Partnering will certainly be the key to the *holistic* approach which must first be brought to the organisation and then incorporated into the team performance with other sub-contractors and the main contractor.' We find no fault with these sentiments, however most of the other current construction management concepts would also subscribe to similar sentiments. The result of this convergence of ideals is that many construction management concepts appear to be in competition with one another for the attention of the same decision-makers.

This can be illustrated in Figure 8.1 by reference to the 'cost influence curve' (based on the Pareto Principle) which has been used extensively in construction management literature. The Pareto principle proposes that the earlier that an individual or group is involved in the decision making process, the greater the potential for impact on the project outcome. Conversely the ability to influence the project outcome diminishes exponentially over time. The problem which we identify is the conflict which can arise when a large number of concepts compete with one another for the prime position at the origin of the x, y axis.

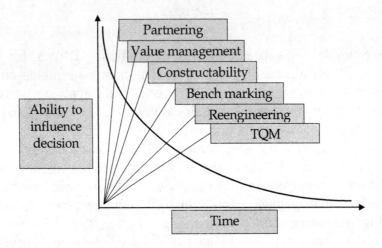

Figure 8.1 Conflicting demands at origin of the cost influence curve

The argument could be made that although government agencies are encouraging and, in some cases, attempting to enforce their adoption of the concepts, no guidance is being given on how the concepts should be applied concurrently and/or in combination.

Rather surprisingly despite arguing that most modern management concepts are underpinned by systems theory, this has not resulted in a unification of the concepts, rather the opposite. The problem lies however, not so much with systems theory, but in the way that it has been applied.

Many researchers in the field of construction management have advocated a systems approach. For example Kelly and Male[10] recommend the use of systems theory and systems thinking in the field of value management as do Chen and McGeorge[11] in the development of a constructability model. It can also be demonstrated that the systems approach underpins reengineering. Checkland[12] however wryly comments 'for some years now a systems approach has been a modish phrase. Few are prepared publicly to proclaim that they do not adopt it in their work and it would be an unwise author of a management science text who failed to sub-title his book: 'a systems approach'. There is much ambiguity about what the systems approach actually is. Often a systems approach is taken to simply imply a holistic view'. Checkland observes however that 'the systems paradigm is concerned with wholes and their properties. It is holistic, but not in the usual vulgar sense of taking in the whole; systems concepts are concerned with wholes and their hierarchical arrangement rather than *the* whole.' The interest for us lies in the notion of hierarchical relationships of concepts.

Although it is evident that certain aspects of concepts such as reengineering and partnering have a certain commonality of purpose, this does not necessarily mean that they are amenable to pigeon holing into some form of hierarchical arrangement. By way of example, both reengineering and partnering have to do with cultural change[13, 14, 8], with the breaking down of existing barriers, with the creation of different sets of relationships and lines of communications. In theory, at least, the concepts are not mutually exclusive and it is possible that reengineering could take place in a partnering environment or that partnering relationships could result from the implementation of reengineering. The relationship is however complex and is likely to be non-hierarchical. Since partnering could be 'nested' inside reengineering or vice versa, if we then start to add successive overlays of concepts such as constructability and total quality management, then, whilst none of them are necessarily mutually exclusive, a conceptual model of the inter-relationships become increasingly complex as successive concepts are added.

There is an intriguing universality about this set of relationships which, as we have previously stated, crosses cultural divides.

Emerging issues

The Pareto concept can also be extended to illustrate the increasing influence of the client in the procurement process and how this is making a fundamental impact on the structure of both the Western and Chinese construction industries.

Figure 8.2 shows different contractual relationships superimposed on the Pareto influence curve, and demonstrates how contractual arrangements and nature of the procurement process has changed over time in response to increasing client demands and increasing stakeholder involvement.

The era from the end of the second world war until the mid-eighties could perhaps be described as the halcyon years for the architect as leader of the traditional procurement process, operating as the leader of the design team with the contractor's involvement being restricted to the post tender process. The mid-eighties and early nineties saw the emergence of the package type process with the advent of design and build, where government clients, in particular were attracted by the risk allocation opportunities of Design and Build. The nineties heralded the emergence Build, Own, Operate, Transfer (BOOT or BOT) with its strong emphasis on stakeholder involvement and full client and contractor participation in the project outcome.

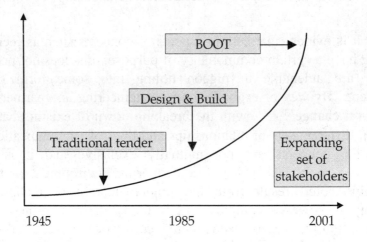

Figure 8.2 Exponential increase in stakeholders' involvement in procurement process

Figure 8.3 shows the ascendancy of professions in terms of client influence plotted on the Pareto curve. The inherent logic of the Pareto influence curve is that in any complex decision making process the players with early involvement in the process have the potential (and opportunity) to make the maximum impact. It is interesting to see how this maxim holds good in terms of professional influence in the post war era, with the profession of Facilities management emerging as one of the dominant players of the nineties. Facility management has been defined by the Centre for Facilities Management[15] as 'the process by which an organisation delivers and sustains support services in a quality environment to meet strategic needs.' Or to put it another way, the provision of a built environment which is aligned to an organisation's goals and objectives. The Total Facilities Management (TFM) approach is most apposite in terms of modern business corporations where there is a strong realisation of the impact of the physical environment on organisational goals and objectives. It is easy to see from this definition why facility management is the fastest growing sector of the construction industry in Europe, USA and Australia. Our observations are that this trend is likely to be replicated in China.

As we have tried to illustrate in Figures 8.2 and 8.3 increasing client involvement and user participation in the provision and performance of buildings is likely to lead to an ever burgeoning and increasingly complex set of relationships between the users and producers of the built environment. For the present, facilities management is likely to be the generic discipline which can best exploit the potential of concepts such as partnering, constructability, total quality management, bench-marking, value management and reengineering. However, what is not yet clear is how to harness the collective power of these approaches, nor how to adopt, adapt or discard these concepts depending on the cultural setting. Perhaps one fundamental aspect (or flaw) which is common to the way in which these concepts have been currently deployed by the industry is that the client has not been fully integrated into the procurement process. Take, for example, partnering although this approach has the capacity to include all the stakeholders, this rarely, if ever happens. More often than not the client, although involved, is to some extent slightly removed from the process. Although, for example, DPWS in Australia has produced a form of contract (C21) which can be used specifically in the partnering process, there is still a sense of an 'us and them situation' between the client and the constructors. Even in the BOOT process there is still to some extent a sense of detachment on the part of the client, given that the client (usually a government authority) only takes possession of the facility at the end of the operating period. This means that in some cases the operating period can extend to between 40 to 50 years (as is the case in the recently constructed Sydney Eastern Distributor tollway).

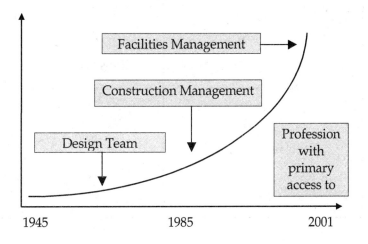

Figure 8.3 The changing roles of the professions

Future directions

One interesting development in terms of total involvement in risk sharing between client and developer is the increasing popularity of alliances. Doz & Hammel[16] have described the growth of alliancing as one of the most dynamic features of modern corporate development *"Scarcely a day goes by without some significant new linkage being announced*". These alliances are proving to be the prevailing sign of the flexible organisation and are a direct response to the changing corporate environment[17]. According to the Economist Intelligence Unit (EIU) alliances of all kinds will rank as one of the most significant management tools by the year 2010. Involvement in alliances is currently cited by 29% of EIU respondents, moreover this figure is predicted to escalate rapidly to no less than 63% by 2010[18].

The alliance process has been described by Gerybadze[19] as "...*the client* (our italics) and associated firms will join forces for a specific project, but will remain legally independent organisations. Ownership and management of the cooperating firms will not be fully integrated although the risk of the project is shared by all participants."

Perhaps the essence of alliancing can best be demonstrated by a description of a typical financial arrangement under a project alliance arrangement. A particular feature of alliancing relationships is the gainshare/painshare approach. This is illustrated in Figure 8.4.

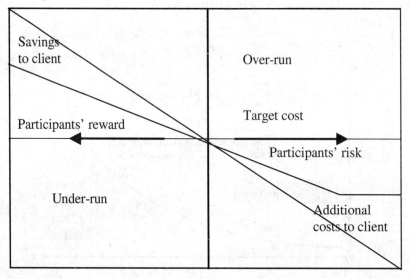

Figure 8.4 Gainshare/painshare alliancing agreement

Gainshare/Painshare

Under this alliancing agreement the gainshare/painshare scheme places as risk, the aggregate of the profit and corporate overhead components of all parties, based on the overall performance of the Alliance against the final cost. In the event of a cost overrun, the participants were to contribute 50% up to a maximum of the total of this profit and corporate overhead, but in the event of a cost saving, 50% of such saving was to be paid to the participants – with no cap, because there was no wish to diminish any incentive. No other penalty provisions were included because the client understood that loss of gainshare, profit and overhead as well as the contribution to painshare, were adequate incentives for performance. The gainshare/painshare split among the parties was agreed based on a 50% allocation to the client with the remainder divided in proportion to each party's contribution in the overall target cost. If the project was completed at less than target cost then additional profits flow to the participants in proportion to their gainshare/painshare percentage. If the project over-ran the target cost then all the participants, including the client are liable for over-run in proportion to their gainshare/painshare percentage.

In our view, alliancing is an interesting combination of the market force culture of profit incentives combined with the cultural ethos of partnering and we believe this could be of particular relevance in the Chinese construction industry in its attempts to use market forces to improve time, cost and quality and achieve the best outcome for stakeholders.

Conclusion

Rigby in his 'Secret history of process re-engineering' posits the notion that all management concepts go through a sequence of six stages viz. deficiency of previous concepts; discovery or rediscovery of a solution; euphoria as early success stories are publicised; over-extension due to the excessive application of the technique to inappropriate situations; derision as examples of failure grow too large to ignore; final abandonment as the technique is discarded or replaced with a new technique. In our attempts to chart current and future issues in construction management and economics we have tried to avoid this level of cynicism, however we are aware that we have probably exposed more problems than solutions. The last decades have seen notable changes in the structure of organisations (both Western and Chinese) as well as methods that managers' use in leading and directing these evolving organisational bodies. At the outset of this chapter we described the introduction of the TCM as an interesting and very innovative response to a difficult and entrenched problem in the Chinese construction industry. We have also discussed cultural shifts in Western approaches to management and some of the difficulties involved in reconciling the current raft of competing management concepts.

We conclude by suggesting that the unpredictable social, economic technical and political aspects of a globalising society will force organisations to look at forming national and trans-national business alliances and we have touched on the emergence of alliancing as one way ahead. Certainly from our exposure to both Western and Chinese construction environments, we believe that both parties have much to gain and learn from each other.

[1] McGeorge D and Palmer A. (1997) *Construction Management: New Directions*. Blackwells Science, Oxford.

[2] Holtham C. (1994) Business process re-engineering: contrasting what it is with what it is not. In: Business process reengineering: Myth and reality editor Coulson-Thomas. 166-173, Kogan Page, London.

[3] Cowan C., Gray C and Larson (1992) Project partnering. *Project Management Journal*, Dec; 5-21.

[4] Hammer M. (1990) Re-engineering work: Don't automate, obliterate. *Harvard Business Review*, July/August; 104-112.

[5] Latham M. (1994) *Constructing the team*. HMSO, London.

[6] Gyles RV. (1992) Royal commission into productivity in the building industry in New South Wales: Government of New South Wales. Sydney, Australia.

[7] New South Wales, Department of Public Works and Services (DPWS). (1995) *Contractor accreditation scheme to encourage reform and best practice in the construction industry*. Sydney: Government of New South Wales, Australia.

[8] Fong P.S.W. (1996) VE in construction: a survey of clients' attitudes in Hong Kong. *Proceedings of the Society of American Value Engineers International Conference*; Vol. 31.0.

[9] Hellard RB. (1995) *Project partnering: principle and practice*. Thomas Telford, London.

[10] Kelly J. and Male S. (1993) Value Management in Design and Construction - the Economic Management of Projects. E&FN Spon, London.

[11] Chen S.E. and McGeorge D. (1993/94) A systems approach to managing buildability, *Australian Institute of Building Papers*, Vol. 5.

[12] Checkland P. (1981) *Systems thinking, systems practice*. John Wiley, Chichester.

[13] Coulson-Thomas C.J. (1994) *Implementing re-engineering*. In: *Business process reengineering: Myth and reality* editor Coulson-Thomas. 105-126, Kogan Page, London.

[14] Kelada J.N. (1994) Is re-engineering replacing total quality? *Quality in progress,* Dec; 79-83.

[15] Alexander K. (1996) *Facilities Management : theory and practice*. 1st ed. E & FN Spon, London.

[16] Doz Y.L. and Hammel G. (1998) Alliance Advantage; the Art of Creating Value through Partnering. Harvard University Press, Harvard.

[17] De la Seirra M. (1995) *Managing Global Alliances: Key Steps for Successful Collaboration*, Addison-Wesley Publishing Company, New York.

[18] Economist Intelligence Unit (1997) *Vision 2010; Designing Tomorrow's Organisation*. Roland Services. New York.

[19] Gameson, R., Chen, S.E., McGeorge, D. & Elliot, T. (2000) Principles, practice and performance of project alliancing - Reflecting on the Wandoo B Development Project. *Proceedings of IRNOP IV Conference*.

BIBLIOGRAPHY

Benchmarking

Camp R. C. (1989) *Benchmarking: The search for industry best practices that lead to superior performance* ASQC quality press, Milewaukee, Wisconsin.

The first and generally viewed as the definite book on benchmarking by the author of the inventor of the technique. It is geared towards manufacturing but its coverage of the topic is extensive.

Spendolini M. J. (1992) *The Benchmarking book*. American Management Association, New York.

Probably the 'best of the rest' of the books on benchmarking, it is more academic than Camp's work and is more comprehensive, having the benefit of being written later.

Codling S. (1992) *Best practice benchmarking: The management guide to successful implementation*. Industrial Newsletters Limited.

This is typical of the other texts listed in the references in the benchmarking chapter. The works do not appear to add much to the subject but are merely the same information presented in different formats.

Peters G. (1994) *Benchmarking Customer services*. Financial Times Pitman Publishing, London.

This work is slightly different in that it concentrates on only one aspect of benchmarking namely customer focus.

Reengineering

Construction Industry Development Agency (1994) *Two steps forward and one step back: management practices in the Australian construction industry*: CIDA, Commonwealth of Australia Publication, Febraury.

Gives good contextual background on the construction culture In Australia.

Coulson-Thomas C.J. (editor) (1994) *Business process reengineering: Myth and reality*.: Kogan Page, London.

Drawing on a detailed, pan-European study of the experience, practice and implications of BPR, the book examines: the advantages and disadvantages of the technique as a tool for change; approaches and methodologies being used; success factors and implementation issues and the implications for organisations and those who work in them.

Hammer M. and Champy. J. (1993) *Reengineering the corporation: A manifesto for business revolution*. Nicholas Brealey, London.

Co-authored by Michael Hammer, who is generally credited with initiating the reengineering movement. Practical advice, including case studies on businesses process reengineering.

MacDonald J. (1995) *Understanding business process reengineering*. Hodder & Stoughton, London.

Explains the challenges and pitfalls surrounding building process engineering and provides a step-by-step guide to understanding the BPR process.

Morris D. and Brandon J. (1993) *Re-engineering your business*. McGraw-Hill, New York.

Introduces the techniques of business process reengineering which brings about the changes in the very processes, practices, assumptions and structures upon which companies are built. It offers guidance on: how modelling and simulation techniques can be used to analyse current operations; the design of a new organisational structure that positions the company to take advantage of changes in the market-place; and the implementation of the new structure.

Petrozzo D.P. and Stepper J.C. (1994) *Successful reengineering*. Van Nostrand Reinhold, New York.

Details how to implement a reengineering program and what to avoid in the process. It stresses the evolutionary application of reengineering business processes, organisations and information systems. The text is divided into four phases of a reengineering project in the order in which they will be carried out. The coverage within these phases includes: selecting the reengineering team; setting the project scope; understanding the current process and information architecture; preparing for redesign; redesign principles; and the final stage of reorganising, retraining and retooling.

Partnering

Bennett J. and Jayes S. (1995) *Trusting the team: the best practice guide to partnering in construction.* Reading Construction Forum, Reading.

Particularly suited to industry practitioners about to embark on partnering or for those who wish to know more about the topic. Covers all aspects of partnering making a clear distinction between project and strategic partnering illustrated by the use of case study material.

Construction Industry Development Agency (CIDA). (1993) *Partnering: A strategy for excellence.* CIDA/Master Builders Australia.

Written by experienced partnering facilitators. Succinctly lays down the principles of partnering, includes pro formas fro partnering charters, action plans, issue escalation process, performance objective evaluation system and rating forms for partnering exercises.

Construction Industry Institute Australia (CIIA) (1996). *Partnering: Models for success.* Research Report 8. Construction Industry Institute Australia.

A study of sixteen partnered projects and two strategic alliances across Australia giving a balanced view of the benefits and disbenefits of partnering, with detailed coverage of the legal implications of partnering in an Australian context.

Cowan C., Gray C. and Larson. E. (1992) Project partnering . *Project Management Journal*, Dec; 5-21.

A journal article co-authored by Charles Cowan, one of the accepted gurus of the partnering movement giving the basic principle of partnering.

Godfrey K.A. Jr., (editor) (1996) *Partnering in design and construction.* McGraw-Hill, New York.

A recent North American perspective on partnering which stresses the relationships between contractor subcontractor and owner.

Hellard R.B. (1995) *Project partnering: principle and practice.* Thomas Telford, London.

Lays particular stress on the need for cultural change and also covers partnering in Australia and New Zealand.

Schultzel H.J. and Unruh V.P. (1996) *Successful partnering: Fundamentals for project owners and contractors.* John Wiley, New York.

A recent North American perspective on partnering authored by two senior industry figures with origins in the Bechtel Company, a partnering pioneer.

Value Management

Miles, Larry. (1967) *Techniques of Value Analysis and Value Engineering* McGraw Hill, New York.

The first book to be written on value engineering and very much based on manufacturing. The text is useful in that despite the developments that have taken place Miles concentrates on value management as a concept as opposed to a technique. At times it appears that the technique has actually come full circle.

Dell'Isola Al. (1988) *Value engineering in the construction industry.* Smith Hinchman and Grylls Washington DC.

A very practical text based primarily on the American 40 hour workshop approach. A lot of advise on how to do value engineering but thin on academic content.

Kelly J.R. and Male S.P. (1993) *Value management in design and construction - the Economic management of projects.* E&F Spon, London.

The only British text on value management in construction, this has a reasonable balance of academic and practical advice on value management. It includes methods of evaluation and a chapter on life cycle costing.

Constructability

Adams S. (1989) *Practical Buildability.* Butterworths, London.

A UK perspective of constructability with case study material dealing with the relationship of design to construction.

Ferguson I. (1989) *Buildability in practice*. Mitchell, London.

A similar work to Adams also with a range of detailed case study material.

Griffith A. and Sidwell A.C. (1995) *Constructability in building and engineering projects*. MacMillan Press, London.

A recent work which comprehensively covers constructability in UK, North America and Australia. Traces constructability from its origins in the Construction Industry Institute (CII) and explains the theory of the Construction Industry Institute Australia (CIIA) Principles File together with a series of case studies illustrating the application these principles over a range of situations from strategic to detailed.

Total Quality Management

Vincent K O. and Joel R. (1995) *Principles of total quality management* Kogan Pages, London.

This text deals with the management of quality, process and quality tools, criteria for quality programmes, and case studies in quality. It also includes chapters on benchmarking and reengineering.

Evans J. R. and Lindsay W. (1993) *The management and control of quality* West Publishing Company, Minneapolis.

Covers similar material to the above but also at the end of each section gives review questions and practice problems.

Asher J.M. (1992) *Implementing TQM in small and medium sized companies*. TQM Practitioner series. Technical Communications Ltd, Letchworth, Hertfordshire.

Despite some information on managing quality in the medium sized organisation the book contains fairly limited information on TQM.

Bounds G., Yorks L., Adams M. and Ranney G. (1994) *Total quality management: Towards the emerging paradigm*. McGraw Hill, New York.

An extremely comprehensive text that deals with both theory and practice of TQM.

Baden H. R. (1993) *Total quality in construction projects Achieving Profitability with customer satisfaction* Thomas Telford, London.

A text aimed purely at the construction industry. A significant portion of the text is devoted to auditing.

Choppin, J. (1991) *Quality through People. A blueprint for proactive quality management.* IFS Publications, Bradford, UK.

This text deals with the people element of TQM but also has sections on the management of the process and of resources. It also presents a blueprint for development of a quality programme.

Imai M. (1986) *Kaizen. The key to Japans competitive success.* McGraw-Hill, New York.

A text devoted exclusively to Kaizen and the Japanese art of management.